CELEBRATIONS OF LIFE

BOOKS BY RENÉ DUBOS

The Bacterial Cell
Bacterial and Mycotic
  Infections of Man
The White Plague
  (with Jean Dubos)
Biochemical Determinants
  of Microbial Diseases
Mirage of Health
Pasteur—Free Lance of Science
Pasteur and Modern Science
The Dreams of Reason
The Torch of Life
Unseen World
Health and Disease
  (with Maya Pines)
Man Adapting
Man, Medicine and Environment
So Human an Animal
Reason Awake
A God Within
Only One Earth
  (with Barbara Ward)
Beast or Angel
Of Human Diversity
Louis Pasteur
The Professor, the Institute
  and DNA
Quest
  (with Jean-Paul Escande)
The Wooing of Earth
Celebrations of Life

CELEBRATIONS OF LIFE

BY
RENÉ DUBOS

McGraw-Hill Book Company
New York St. Louis San Francisco Bogotá Guatemala
Hamburg Lisbon Madrid Mexico Montreal Panama
Paris San Juan São Paulo Tokyo Toronto

First paperback edition, 1982

1 2 3 4 5 6 7 8 9 FGFG 8 7 6 5 4 3 2

ISBN 0-07-017894-1

LIBRARY OF CONGRESS CATALOGING IN PUBLICATION DATA

Dubos, René Jules, 1901–
Celebrations of life.
1. Man. 2. Human evolution. I. Title.
GN27.D8 302 81-3764
ISBN 0-07-017893-3 AACR2
0-07-017894-1 (pbk.)
Book design by Jerry Wilke

FOREWORD ix

1

THE HUMANIZATION OF HOMO SAPIENS 3

THE GLOBAL HUMAN FAMILY
MY ROOTS
HUMAN BEINGS AS ANIMALS
SOCIAL DIVERSITY WITHIN HUMANKIND
HOMO SAPIENS INTO HUMAN BEING
THE ORIGINS OF HUMANKIND
THE SOCIALIZATION OF *HOMO SAPIENS*
THE INVARIANTS OF HUMANKIND

2

THE PAST, PUBLIC PLACES AND SELF-DISCOVERY 37

LIFE AS EXPERIENCE
SURVIVAL OF THE PAST
NATURAL AND BUILT ENVIRONMENTS
IMAGES OF HUMANKIND
CHOICES AND CREATIVITY
SELF-DISCOVERY

3

THINK GLOBALLY, BUT ACT LOCALLY 83

LOCAL SOLUTIONS TO GLOBAL PROBLEMS
THE GLOBAL VILLAGE
THE NETHERLANDS, HORIZONTAL COUNTRY CREATED BY HUMANKIND
MANHATTAN, THE VERTICAL CITY
THE GLOBE-TROTTER AT HOME

4

TREND IS NOT DESTINY 131

THE BEAUVAIS SYNDROME
THE MATERIAL POWER OF SPIRITUAL FORCES
WORLD TRENDS AND CONTEMPORARY GLOOM
SOCIAL ADAPTATIONS TO THE FUTURE

5

MATERIAL RESOURCES AND THE RESOURCEFULNESS OF LIFE 155

FROM THE WILDERNESS TO HUMANIZED NATURE
RAW MATERIALS AND RESOURCES
THE MERITS OF ENERGY SHORTAGES
CREATIVE ADAPTATIONS AND ASSOCIATIONS

6
OPTIMISM, DESPITE IT ALL 195

CIVILIZATION AND CIVILITY
THE SEARCH FOR CERTAINTIES
HUMAN COMMUNITIES
WEALTH, TECHNOLOGY AND HAPPINESS
SOCIAL PRIORITIES
DAYDREAMING ABOUT THE FUTURE

ENVOI 253

INDEX 254

FOREWORD

My first intention had been to entitle this book *The Celebration of Life.* On second thought, however, Celebrations seemed more appropriate because there are as many ways of experiencing the world as there are living creatures. Who can doubt that a cat stretching itself in the sun or curled near the warmth of a stove, and that a lamb or a colt frolicking in a spring meadow, are celebrating life each in its own way. We human beings can supplement this purely biological joie de vivre by inventing life-styles, social institutions, patterns of thought and innumerable other activities that transcend biological needs and constitute as many different ways of experiencing and celebrating human life.

The word "life" denotes not what living organisms are made of, but what they *do.* Observations and scientific studies have provided much knowledge about living creatures and especially human beings—their evolutionary development, their anatomical characteristics, their physiological mechanisms, their behavioral patterns—but this biological knowledge does not reveal how life is experienced. We can experience life only through our total being and it is by extrapolation that we endow other human beings with similar experiences. We celebrate life when we acknowledge that it is at the origin of all our satisfactions, and that it gives meaning to our purely biological existence, even when we fail in our enterprises. In Ruskin's words, "There is no Wealth but Life."

This book is not about the nature of life, or about the characteristics or activities of particular organisms, but about the living experience, especially in the human adventure. I shall try to convey this experience by stating what I *know* and how I *feel* about certain places and events that have been of particular interest to me or that have affected some aspects of my own existence.

Chapter I, "The Humanization of *Homo Sapiens,*" documents that members of the biological species become really human and master a human language only if raised until a critical age in a human society, anywhere on earth.

Chapter II, "The Past, Public Places, and Self-Discovery," discusses how a particular human being becomes a person—unique, unprecedented and unrepeatable—as a result of heredity, the effects of surroundings and events, and especially the choices made in the course of existence.

Chapter III, "Think Globally, but Act Locally," illustrates through historical and contemporary examples that, whereas most important problems of life on earth are fundamentally the same everywhere, the solutions to these problems are always conditioned by local circumstances and choices.

Chapter IV, "Trend Is Not Destiny," illustrates that, whereas biological evolution is irreversible, social evolution makes it possible for human societies and individual persons to change their course and even to retrace their steps when they judge they are moving in an undesirable direction.

Chapter V, "Material Resources and the Resourcefulness of Life," describes how human beings, and other living organisms, can convert the raw materials and forces of the earth into substances and structures that they use for their own development, as well as to carry on their activities.

Chapter VI, "Optimism, Despite It All," affirms that despite past and present tragedies, we can have faith in the future because human life and nature are extremely resilient and because we are learning to anticipate the dangers inherent in natural forces and in our own activities. The future cannot be predicted, but we are still renewing ourselves as we move on to new places and new experiences—the human way of celebrating life.

Celebrations of Life is an expression of my personal experiences and I am indebted to the countless persons who have helped me throughout the more than eighty years of my existence. Since I could not possibly thank them all individually, I shall mention only three who symbolize three stages and aspects of my life.

Adeline De Bloëdt Dubos—my mother, to whom I owe everything.

Jean Porter Dubos—my wife, who has contributed creative ideas and criticisms to every part of this book.

Peggy Tsukahira, who first discussed the book with me and who looked so sad after she had read parts of the first version that I immediately reorganized the manuscript in an attempt to make the text more readable and to bring back her lovely smile.

While I am responsible for every phrase in *Celebrations of Life,* including the title, I should mention that many of its concepts originated during discussions with friends and colleagues who have created the René Dubos Center for Human Environments, in particular with Ruth and William Eblen. The activities of the Center are based on our faith that it is possible for human beings to modify the surface of the earth in such a way as to create environments that are ecologically viable, esthetically pleasurable and economically profitable, generating thereby even more opportunities for the celebration of life.

1

THE HUMANIZATION OF HOMO SAPIENS

THE GLOBAL HUMAN FAMILY

MY ROOTS

HUMAN BEINGS AS ANIMALS

SOCIAL DIVERSITY WITHIN HUMANKIND

HOMO SAPIENS INTO HUMAN BEING

THE ORIGINS OF HUMANKIND

THE SOCIALIZATION OF *HOMO SAPIENS*

THE INVARIANTS OF HUMANKIND

1

THE HUMANIZATION OF HOMO SAPIENS

THE GLOBAL HUMAN FAMILY

After three weeks of strenuous activities under the sparklingly luminous spring skies of Australia, followed by a week in the hot, humid and polluted air of Taipei in Taiwan, my wife and I decided to stop in the Hong Kong Territory before returning to New York. This was in 1965. Hong Kong was then, and still is, the metropolis with the highest population density in the world; the average income was extremely low, especially among the refugees from continental China. Surprising as it may seem, however, all medical and social surveys that I had read described the states of physical and mental health as excellent throughout most of the city, and reported that crime or other forms of disorderly conduct were rare. We were eager to see whether seemingly pleasant ways of life can be really achieved under conditions of extreme crowding and poverty.

We did not know anyone in the Hong Kong Territory nor anything about its structure. As we wanted first to explore the inner part of the city without being influenced by guides, we walked more or less at random, responding only to the intriguing stimuli that reached our eyes, ears, and noses.

We saw everywhere and at all times of the day people eating either in the street or in a great variety of restaurants located on boats near the shores as well as on the land. Immense numbers of healthy children were in a playful mood yet behaved in an orderly fashion as they spilled out of schools of various grades and specialties. In the evening we walked through very poor sections of the city which nevertheless did not seem to qualify as slums because the streets were used for an endless diversity of lively and cheerful spectacles.

Even though we were the only Caucasians in most situations, we did not experience at any time concern for our safety or even embarrassment. High population density and low income thus did seem compatible with a pleasurable and decent life. It is probable that crowding becomes a public danger only when other social conditions have seriously deteriorated.

In midafternoon we ventured into a huge complex building, part public market and part department store, which turned out to be operated by the People's Republic of China. While surveying the displays, we noticed an adorable little Chinese girl engaged in a lively conversation with an elderly Chinese gentleman. Tall with a thin face and a pointed white beard, he looked extremely wise and distinguished in his plain but very neat clothing. He and the little girl seemed the perfect symbol of the loving relationship between grandfather and granddaughter anywhere in the world. Both the Chinese gentleman and the little girl were much more interested in each other than in either the displays or the other people, but it was obvious from the way they looked at us that they had become aware of our interest in them and of the pleasure we derived from their presence.

My wife and I moved on through the different sections of the store and, just for the sake of owning an object manufactured in continental China I bought a shirt of a Mao semi-military style. The shirt fitted me well but when I put it on in New York it still had, even after washing, a peculiar odor and aura that was incompatible with my American life. It eventually found its way to a thrift shop.

We continued our walk in the market and some fifteen minutes after my purchase of the shirt we saw again the little girl with her grandfather. The four of us exchanged friendly smiles probably initiated by us. But my wife and I were in a serious mood at that time as we were investigating furniture which was beautifully made yet led us to wonder why chairs that were of the right proportion for our bodies made us feel awkward, as if we had to learn gestures and attitudes new to us, although natural to the Chinese people.

The hour had come when we wanted to find our way back, on foot, to the hotel. As we were about to leave the market our paths crossed once more with the little girl and the old gentleman but on this third encounter both their smiles and ours exploded into peals of hilarious laughter. We had the impression that we had known both of them for a long time and we felt that they, too, knew much about us—indeed everything about us which is of any real importance in human relationships. We, the Chinese grandfather and his granddaughter, had been separated culturally for countless generations and separated geographically by several thousand miles, yet we could immediately relate to each other because we were parts of the same biological group, the global human family.

I have had similar warm contacts that did not depend on verbal communication in other parts of the world; for example, with Indian children and their parents in a New Mexico school while I was working on the Navajo reservation during the late 1940s, and with two Senegalese families while I was stranded for a day at the Dakar Airport in the 1950s. The happiest smile I ever saw was in a film of a little Australian aborigine girl displaying a huge, juicy grub she had just extracted from under the bark of a tree and was about to eat. I understood how happy she felt at that moment and I am convinced that she would have perceived the pleasure I derived from watching her if we had been face to face. Even the most unfamiliar behavior of human beings is meaningful to us for the seemingly obvious, but in reality mysterious, reason that there is a peculiarly characteristic human way of doing almost anything.

Everywhere on earth, all sorts of people are constantly engaged in all sorts of activities which they carry out in countless different ways. Yet an overwhelming impression of unity emerges from this prodigious diversity of human types, social situations and behavioral patterns. An Eskimo harnessing his huskies, a Chinese farmer stooping in a rice paddy, a Tuareg riding his camel through a Sahara dust storm, a New Yorker hailing a taxicab, a Japanese tourist taking photographs in the park of Versailles, all display behavioral patterns that are distinctively human.

The attitudes of scientists concerning human nature have ranged widely during recent decades. In 1928, for example, Margaret Mead published her first book, *Coming of Age in Samoa,* in which she concluded from her studies of the young women living on that island that much of what we ascribe to human nature is in reality the expression of the social environment in which a person is born and raised. In 1978, half a century after the publication of Margaret Mead's *Coming of Age,* Professor E. O. Wilson achieved fame, compounded of both admiration and outrage, by affirming in his book *On Human Nature* that the genetic heritage of the human species governs all aspects of our behavior—not only aggression and sexuality, but also traits assumed to be characteristically human such as generosity, self-sacrifice and even religious feelings.

In *Celebrations of Life,* I shall take it for granted that both the genetic constitution and the total environment play roles in all aspects of human development and behavior, but I shall particularly emphasize that these deterministic mechanisms do not fully account for human life. Persons and societies do not submit passively to surroundings and events. They make choices as to the places where they live and the activities in which they engage—choices based on what they want to be, to do and to become. Furthermore, persons and societies often change their goals and their ways; they can even retrace their steps and start in a new direction if they believe they are on the wrong course. Thus, whereas animal life is prisoner of *biological* evolution which is essentially irreversible, human life has the wonderful freedom of *social* evolution which is rapidly reversible and creative. Wherever human beings are concerned, trend is not destiny.

On several occasions I shall introduce the themes of this book by examples taken from my life, not only the surroundings and events that have influenced my development but also the choices I have made. This will not be to emphasize the circumstances of my life but rather to use them as stepping stones through a wider stream of human experience to other situations in various social and national groups. The following brief outline of my life is meant to convey the general truth that all of us incarnate our environmental and social pasts, and that we cannot escape from their influence on our constitution and our behavior.

MY ROOTS

I was born at the beginning of this century—on February 20, 1901, to be exact—and was raised in agricultural villages some 50 kilometers directly north of Paris. Until my family moved to Paris when I was almost thirteen years old, I had seen only one small town located a few miles from the villages where I grew up. It was Beaumont-sur-Oise where my grandparents lived and which did not exceed 3000 inhabitants. I left France in early 1922, first for Italy, then in late 1924 for the U.S. and have had very little contact since that time with the region in which I was raised. In fact, I did not return to Hénonville, the tiny village in which I spent most of my childhood, until sixty years after I had left it for Paris.

I have never regretted having removed myself from the part of the world in which I was raised, nor have I ever experienced homesickness—

le mal du pays. Yet the names of places associated with my youth always awaken in me deep emotional responses. Merely to pronounce them is sufficient to transport me through time and space. I still derive deep pleasure from writing or speaking names such as *Le Pays de Thelle* and *Le Vexin français,* the two minuscule regions of the Île de France in which my villages are located. It is also a wondrous experience to read the names Oise, the broad sluggish river where I went fishing with my grandfather, and La Troesne, a narrow brook that flows from Hénonville into the Epte River at Clermont en Vexin, and which was the first running stream I ever saw.

How strange that I should respond with such intensity to the names of places which I have experienced only during my early youth and which have played no obvious role in my life since I left them more than half a century ago. At the most, I can rationalize my attitude by remembering that names such as *Vexin* and *Pays de Thelle* evoke the charm and also the limits of provincial life in the past; that the names *Oise* and *Troesne* symbolize the peaceful streams along which people have settled since prehistoric times; that the phrase *Île de France* conjures the diverse and gracious landscapes which have served as stages for the development of the most distinctive aspects of French history and French culture.

Of more direct relevance to me, however, is that the names of places where I spent my youth also call to mind the situations in which I became conscious of my own life and developed my personality. They evoke the countryside as I perceived it, the persons with whom I associated and the activities that I regarded as normal ways of life. The hereditary biological characteristics that I acquired from my parents were only potentialities until they were shaped by the ways of life in the Île de France. They became actual physical and mental attributes that still survive in me despite my long absence and countless subsequent experiences.

Readers who are interested in "roots" will be disappointed by my ignorance concerning my familial past. My father's name Dubos certainly originates from the southwest of France where it is extremely common, but he and his father were born north of Paris. My grandfather Dubos was a house painter in the small town of Beaumont-sur-Oise; at the age of thirty-five he won some money in a national lottery and immediately retired from business. He bought three houses in Beaumont, settled in the best one, and derived income from renting the other two, located less than five minutes walk from his new home.

Although I had many contacts with him as a child, I cannot recall that he had any particular interest except for gardening and for fishing in the Oise River—not from a boat but almost exclusively from the banks. Of his wife, my father's mother, I remember only that she was handsome, strong willed, and that she was the absolute ruler of the household.

My father became a butcher boy in Beaumont and met my mother during his military service in Sedan, a small city in the northeast of France. I had very few contacts with him because all his time was monopolized by the small butcher shops he owned. In 1914 he was conscripted at the beginning of the First World War and died in 1918 while still in uniform.

The account of my mother's ancestry sounds more romantic but has never been documented. According to her parents, her father was found as a very young infant, abandoned in the Belgian cathedral of Saint Gudule in Brussels. He was dressed in fine attire with a note pinned to him, instructing that he be taken to a certain farm near Mons, close to the French border, where money would be provided for his maintenance until he came of age. His name was given as "De Bloëdt," a Flemish word meaning "the blood." He eventually moved to France, settled in Sedan as a worker in a textile factory, married a humble French person and my mother was born as Adeline De Bloedt. She was rather short, dark haired, sensitive, vivacious and, as will become obvious later, had a most profound influence on my life—even though I probably derived my physical appearance less from her than from my father.

Surprising as it may seem, the small villages and the farming countryside in which I grew up have not changed much during the past seventy years. I cannot judge whether their very humanized environments are banal or attractive. I only know that I feel comfortable in them, almost certainly because they have shaped my physical and mental being and I am still adapted to them—even after almost seventy years of absence.

By the age of twenty, I had acquired the essential attributes that had resulted from the interplay between my genetic constitution and my environmental conditioning and that remain the dominant aspects of my personality. I was then almost six feet tall, rather lanky, with blue-green eyes and abundant blond hair—Viking-like. I was hardworking and eager to get on in the world, perhaps chiefly for the sake of adventure—again Viking-like. I would not be surprised if much of my genetic constitution could be traced to some of the Nordic people who settled in France during the Dark Ages. But my voice, my gestures, my tastes and my ways of life were the products of my social conditioning in twentieth-century France and made me very different from what I would have been if my youth had been spent in Scandinavia or in some other part of northern Europe. As far back as I can remember, I have loved walking in the woods and even more so in open fields, but I have rarely experienced in them the romantic imaginings that are said to be common among young people with a Swedish, German or Russian soul.

The existential manifestations of my genetic endowment—call it my biological nature if you will—were largely determined by the exhilarating capriciousness of the Île de France skies—often delicately luminous but at times dark and mildly threatening; by the diversity of the French landscapes, either highly structured and disciplined as on the tree-lined roads and formal gardens, or somewhat disorderly and mysterious as in a Corot painting; by the ordinary people whom I knew who were active and hardworking but occasionally sentimental and usually exuding a lively joie de vivre.

The preceding paragraphs might lead the reader to conclude that I am now an old man, living entirely in the past, but the truth is that I am still very active on the present scene, very much a part of modern times. When I organized the René Dubos Center for Human Environments and dedicated it in October 1980, my concern was not the preservation of landscapes and monuments, but a wish to demonstrate that the best policy of conservation is to intervene carefully yet creatively in the existing order of things.

HUMAN BEINGS AS ANIMALS

We are human not so much because of our appearance, but because of what we do, the way we do it, and more importantly because of what we elect to do or not to do. Our species acquired its humanness not by losing its animal characteristics, but by engaging in activities and developing patterns of behavior that have led to a progressive transcendence of animality resulting in the creation of humanness.

A truly human language and type of behavior has been achieved so far only by the members of the biological species *Homo sapiens* but the conclusion that this makes us qualitatively different from the rest of the animal kingdom has long been questioned and is being increasingly questioned today. As now recognized by all anthropologists, the more we learn about primate behavior, the smaller the difference between human and nonhuman primates appears to be. This view is seemingly in agreement with the recent laboratory findings that the genetic difference between chimpanzees and human beings is less than one percent. There are many people who believe indeed that apes could be taught to behave, talk and even think in a human way if they were placed under the right conditions, and handled from the time of birth as if they were human children.

When Europeans first encountered the great apes a few centuries ago—the chimpanzees in West Africa, then the orangutans in Sumatra and Borneo, and finally the gorillas in central Africa—they were deeply impressed by some of the humanlike attributes of these creatures and regarded them as wild hairy humans with tails. The president of the Berlin Academy of Sciences was much intrigued by the stories of explorers concerning orangutans and stated in 1768 that he would rather spend one hour of conversation with one of these "hairy men" than with the finest minds of Europe. He may have known that the word orangutan is derived from the Malaysian word for wild man.

I should at this point state my own prejudice that real language—whether articulated or not—is one attribute that really differentiates human beings from other animal species. Parrots and mynah birds can learn to speak English words as human infants do, but the learning of the birds is merely mindless mimicry.

It is most unlikely that a real conversation can ever take place between human beings and any other primate. Several primates, of different species, have been raised from youth in human families and given every possible inducement to develop a kind of human language and to behave in a human way. Under these conditions, they have indeed learned to understand and to communicate a few simple concepts. To a limited extent some of them may even have been able to symbolize a few simple aspects of the external world and of their relationship to their mentors, although this is more questionable. But none has succeeded in acquiring a really human language or behavior. In fact, this may be an unreasonable expectation. Even if *we* could master the techniques of communication that apes or other animals use among themselves, we would have to change our own social nature before we could think of topics that would lead to a real conversation. Whereas language is chiefly if not only a means of communication among animals, among human beings it is a mechanism for the formulation and use of symbols that are basic to the creativeness of human life.

I hasten to acknowledge that human beings have probably always realized—at least since Cro-Magnon time—that they have profound affinities with the other members of the animal kingdom. For example, many hunter societies, however primitive, seem to have shown signs of respectful behavior toward animals they hunted although they often killed many more animals than was justified by their needs. For thousands of years, furthermore, the composers of bestiaries have used portraits of animals to symbolize religious teachings and the ethical problems of human life. With time, the themes of bestiaries became increasingly secular and animals were used to represent the various aspects of ordinary human behavior on earth. *Le Bestiaire d'Amor* of the thirteenth-century illustrates the games of love among humans. Dante used animals in the *Divine Comedy* to symbolize the vices, passions and virtues of his time. Machiavelli selected the lion and the fox to convey in *The Prince* the roles played by strength and cunning in political conflicts.

The popular allegories based on animal life may have lost some of their appeal after Descartes asserted in his philosophical treatises that animals are nothing but machines, but this eclipse, if it really did occur, was never complete and did not last long. In fables, those of La Fontaine, for example, animals soon recovered their symbolic significance in the representation of the tragedies and comedies of the human condition, with all their nuances. The French philosopher Hippolyte Taine devoted many pages to their symbolic value in his book, *La Fontaine et ses Fables*. In our times, the successes of Orwell's *Animal Farm* and of Bach's *Jonathan Livingston Seagull* demonstrate that modern people still acknowledge affinities with animals, at least to the extent of representing certain types of social behavior. Among all people of the world, indeed, there has always been a tendency to personify by one or another animal species some particular aspect of human nature—intelligence or stupidity, aggressiveness or timidity, extravagance or frugality, carelessness or prudence.

The representation of human life by animals has its counterpart in the all but universal habit of describing animal behavior with words derived from human behavior—as if animals conducted their lives according to a repertory of attitudes similar to that characteristic of human beings. In the eighteenth-century, this practice was widely followed by Buffon in his *Histoire Naturelle;* in our century it was used by Sir Julian Huxley in his early writings on animal ethology. For example, the word "ritualization" is heavily loaded with anthropomorphic and historical values, yet Huxley adopted it to denote the symbolic confrontations by which animals try to achieve dominance over other animals of the same species within a given group, as if the symbolic combats among male wolves were equivalent to the jousting of medieval knights.

While human beings have always acknowledged their affinities with animals, the immense majority of us have considered human life superior to animal life by virtue of certain intellectual and moral attributes regarded as peculiar to humankind, but attitudes seem to be changing among our contemporaries in this respect. The new intellectual fashion is to pretend that the human species barely differs from other animal species and to explain human behavior and history by a purely biological determinism.

John Locke, Jean Jacques Rousseau and other partisans of the "nurture" theory of human development taught that the newborn child is like a blank page on which everything is consecutively written by experience and learning in the course of life. A century ago, Thomas Huxley gave a more biological expression to this thesis when he asserted that the newborn infant "does not come into the world labelled scavenger or shopkeeper or bishop or duke, but is born as a mass of rather undifferentiated red pulp" the potentialities of which can be revealed only by education. The "nurture" theory has taken many forms in our times. Sigmund Freud and his followers believed that the peculiarities of each person's mind can be accounted for by early influences, including and especially those around the time of birth. Most of the difficulties that plague human existence would thus be determined by the early environment. This was Margaret Mead's general view, and that of Columbia University's School of Social Anthropology of which she was a part when she gained fame with her book, *Coming of Age in Samoa.* At the present time, B. F. Skinner still holds the most extreme position with regard to the effect of the environment on behavioral determinism, as illustrated by his laboratory techniques for the conditioning of experimental animals, and by his books—such as *Walden II* or *Beyond Freedom and Dignity*—where he assures us that we could create any type of social behavior that we wish by shaping the proper kind of social environment for humankind. Many are those who believe that this is being done now by the Madison Avenue public relations specialists.

On the other hand Carl Jung's writings early in the twentieth-century implied that humankind can be understood only by exploring the many factors which played a part in the genesis of the human mind during the remote past; according to him, much of individual behavior is influenced by archetypes as old as the human race itself.

Contemporary advocates of genetic determinism are more specific. In their views, we are nothing but naked apes; our relationships with other human beings and with the rest of the cosmos are governed by territorial imperatives and other aggressive and even destructive attributes inherited from our Stone Age ancestors who had to be "killers" because they derived their living from the hunt. According to Professor Edward O. Wilson, the present leader of this school of sociobiology, even altruism and religious feelings are the consequence of genetic mechanisms that once had and usually still have a selective survival value.

Professor Wilson has no difficulty in proving that, in the case of social insects, "natural selection has been broadened to include kin selection." The termite soldier, for example, protects the rest of its colony by its self-sacrifice, with the result that its more fertile brothers and sisters flourish. The argument, however, is much more complex in the case of human beings, and can be illustrated only by unconvincing examples. I shall therefore limit myself to quoting from the beginning and the end of the long chapter that Professor Wilson devotes to altruism in his book, *Human Nature.* "The blood of martyrs is the seed of the church. With that chilling dictum the third-century theologian Tertullian confessed . . . that the purpose of sacrifice is to raise one human group over another. . . . Human behavior . . . is the circuitous technique by which human genetic material has been and will be kept intact. Morality has no other demonstrable ultimate function." According to Wilson religious practices also confer biological advantages because, in the midst of the potentially disorienting experiences of each person in daily life, religion leads to membership in a group and thereby provides a driving purpose in life compatible with self-interest.

Although behaviorism and sociobiology are scientifically poles apart in the biological mechanisms they invoke, they derive from a similar attitude with regard to human life. With either explanation, the human being loses identity as *subject* since it is shaped and governed by forces over which it has no control. The person becomes a mere *object,* whose behavior and fate do not involve conscious choices. Human beings become products of "chance and necessity" for whom freedom and dignity are just meaningless concepts.

It needs hardly to be stated once more that *Homo sapiens* is an animal very similar to the primates in anatomical structure, in the physiological mechanisms that keep the body machine going, and in the instinctive responses it makes to environmental stimuli. But knowledge of these animal aspects of our nature is not sufficient to account for our humanity. This is what the German psychologist Benno Erdmann had in mind when he wrote almost a century ago, "In my youth, we used to ask ourselves anxiously: What is man? Today scientists seem to be satisfied with the answer that he *was* an ape." More recently, Paul Valéry also expressed dissatisfaction with the orthodox scientific explanation of human life. His statement, "L'homme n'est pas si simple qu'il suffise de le rabaisser pour le comprendre," means that humanness cannot be accounted for only in terms of biological structures and mechanisms. Even Émile Zola, the champion of the "scientific" novel, stated in the notes for his novel, *La Joie de Vivre,* "I want to show my hero trying to reach happiness by struggling *against* his innate hereditary characteristics and *against* the influences of his environment." In the same spirit, I shall focus my attention in the following pages on the characteristics that differentiate human beings from animals. I shall emphasize in particular that these distinctive characteristics do not derive from the *biological* attributes of the species *Homo sapiens* but from the *manifestations* of human life which are to a large extent the consequences of intentionality.

SOCIAL DIVERSITY WITHIN HUMANKIND

Most if not all of the social traits found in one or another human society are also found in the life of one or another species of primate. For example, the social organization among nonhuman primates includes practically all possible kinds of sexual associations, from the solitariness of the orangutan to the gregariousness of the chimpanzee; from the monogamy of the gibbon to the polygamy of the baboon; from the seminuclear family of the gibbon to the loose associations of the baboon and chimpanzee; from male-female cooperation and equality among gibbons to male dominance and separation between the sexes among baboons, chimpanzees, and gorillas. All these varieties of sexual and social arrangements are found among human beings, and a similar diversity exists for all other types of human organizations or activities. In consequence, intentionality and freedom of choice are at least as important in human life as is biological determinism—whether genetic or environmental. To a very large extent, human beings can overcome the constraints created by their genetic peculiarities and by the environments in which they live and function; they usually have a good deal of freedom in selecting these environments and their ways of life.

For example, although human beings are assumed to have reached the American continent from Asia as far back as 20,000 years B.C., and perhaps even by 50,000 years B.C., the social patterns of the various groups of Amerindians encountered by the Europeans on their arrival in the New World differed profoundly from those of Europe, and even more strikingly from one part of America to another. The first arrivals from Asia brought with them some form of Stone Age culture and a knowledge of how to use fire. In time, some of them learned how to use metals but others never advanced beyond stone weapons and tools. None of the Amerindians used the plow or the wheel and the tongues they spoke were so varied that some tribes separated only by a few miles had to use sign language in order to communicate.

Great civilizations comparable in splendor to those of Europe and Asia developed early in the jungles of Central America, the Valley of Mexico and the Peruvian highlands; but the equivalent of the Maya, the Aztec and the Inca empires did not emerge anywhere in the vast, almost empty sweep of territory that was to become the United States and Canada. Here, in 1492, the Indian population was approximately one million as compared with the estimated fifteen million in Central and South America. Even more striking, however, was the extreme diversity and fragmentation of the North American tribes. They ranged in temperament from the peaceful Pima of Arizona to the belligerent Iroquois of the New York area. Some were town dwellers like the Pueblos of the Rio Grande Valley; others were nomadic and hunters like the Apache. Some, like several Pacific Northwest tribes, had a capitalistic organization whereas many eastern tribes were communally oriented. Even when they lived close to each other, as was the case for the Hopi and the Navajo, they had different languages, ways of life and religious beliefs.

There were probably some five hundred different Indian tribes in pre-Columbian North America, each with its own life-style. The arid Southwest had developed a settled farming tradition, with elaborate networks of irrigation canals in the Gila River–Salt River country. In contrast, the primitive Shoshoni who lived in comparative isolation on the harsh plateau area north of Pueblo derived their subsistence from the hunt and from the gathering of wild seeds and piñon nuts; their social organization rarely rose above the family level. The Indians of the Northwest coast had a capitalistic organization; they were a maritime people but also remarkably skilled in utilizing the stands of towering evergreens. They fished for salmon in the great rivers, plied the coast in their dugout canoes and used their towering totem poles to document history. Their many-layered social organizations ranged from chiefs and nobles to slaves.

The Great Plains east of the Rockies seem to have been fairly quiet before the introduction of the horse by the Spaniards but this serenity was shattered when nomadic tribes like the Comanche, the Apache, the Blackfoot and the Sioux moved in with mustangs and made the grasslands a range for buffalo hunting and also for their tribal wars.

East of the Great Plains were numerous small tribes loosely organized in larger confederacies. In the Algonquian group, men stalked deer and rode the streams and lakes in their birchbark canoes while women tended small plots of corn, squash and beans, the people lived in villages of domed wigwams and were protected by fortifications of tree trunks. The Algonquian Indians needed these mechanisms of protection because they were constantly threatened by other Indians of an aggressive temperament more or less banded together as in the Iroquois confederation. Other warlike tribes with other ways of life, but also linked in weak confederacies, occupied the Southeast. Best known among them are the Natchez who inherited a mound-building tradition that began in the very distant past in the Ohio Valley and that progressively evolved into the building of more elaborate sprawling earthen mounds—a kind of flat-topped pyramid that served as the foundation for temples or palaces. Another Natchez claim to distinction was a complex, class-structured society that included the only absolute monarchy so far recognized among the American Indians.

These examples of extreme diversity among Amerindians are not meant as a presentation of life in pre-Columbus America, but only as an illustration of the fact that biology does not *determine* the social aspects of human behavior. All members of the species *Homo sapiens* are endowed with similar biological and mental potentialities; all of us are limited by similar constraints in what we can do, but in actual life we function to the beat of very different cultural drums. I realize the obviousness of this statement, an expression of simple common sense. But we live in a period when many people, and not only ivory tower scientists, are so intoxicated by recent biological discoveries that they no longer appreciate the extent to which the social diversity of humankind is one of our most distinctive traits. Theoretical biology is easier to grasp than the complexities of human life and therefore encourages the formulation of simplistic dogmatic theories of life. For this reason, it has become necessary to restate certain very elementary truths because they are blurred by the acceptance of incomplete and poorly assimilated scientific knowledge.

HOMO SAPIENS INTO HUMAN BEING

Much confusion comes from the general belief that the expressions human being and *Homo sapiens* have exactly the same meaning, whereas they really differ profoundly in their connotations.

Members of the species *Homo sapiens* are not born with the attributes essential for a truly human life but rather with *potentialities* that enable them to *become* human. These potentialities can be expressed only if the newborn *Homo sapiens* creature has the opportunity, from very early in life, to grow and function among other human beings, in *any one* of the many different kinds of human societies. We become human only to the extent that we take advantage of these opportunities. I emphasize *any one* because past and present experience demonstrates that people of all races and skin colors can rapidly learn to live and function effectively among other people if they have been socialized early in life, even in a very primitive society. People of different social groups think, feel and speak differently about different things, but they can all think, feel and speak—the attributes that transform *Homo sapiens* into a human being.

When stripped to our purely biological basis, we are just animals closely related to the higher apes. Our biological nature cannot account alone for our social patterns and cultural concerns, and even less for the distinctive persona by which each one of us is known, and which we create to a very large extent ourselves through our own choices. The difference between animality and humanness can be illustrated by one of the simplest, yet most striking behavioral difference between animals—even the most noble and spectacular—and human beings—even the most primitive.

In theory, lions, tigers, polar bears, orangutans, gorillas, and other powerful animals could readily extend their habitats by displacing other creatures. But in nature they rarely if ever move out of the natural environment in which they have evolved and to which they are biologically adapted; they even remain highly localized within this environment. The same can be said of practically all other animal species, weak as well as strong. Migratory birds are not an exception to this rule. Far as they may travel, it is along a preordained course and according to a seasonal program to which they must conform. In the wild, the good life for an animal means carrying out those particular activities for which its instincts have been programmed during its evolutionary development in its natural habitat.

The reason for this "parochialism" of wild animals is not that they could not survive under conditions different from those of their native habitats. The zoo experience proves that, with minor accommodations, most species can live and reproduce in places far removed and under conditions very different from the natural environments in which they have evolved. Animals in the extremely popular Central Park Zoo of mid-Manhattan are generally in excellent health; their life span is commonly longer than it would be in the wild and most of them breed successfully. One notable event of New York City life in 1972 was the birth of the gorilla Patty Cake whose mother Lulu displayed an ideal motherly behavior, loving and vigilant, seemingly undisturbed by the presence of countless admiring but noisy and agitated New Yorkers. At the time of writing, November 1980, Patty Cake is alive and well.

Animals stay in their native habitats probably for the simple reason that they have no need to seek conditions other than the narrow range to which they have been adapted biologically and behaviorally by Darwinian evolution and by the accidents of their birth and upbringing. It is doubtful that they can even conceive of existence under conditions other than the ones under which they have developed. And yet, they may respond to the "natural" conditions of their evolutionary past, even if they have never experienced them. I remember the rapidity and intensity with which a house cat directed its eyes toward the upper part of the room when it heard for the first time phonograph records of bird songs, even though it had been born and raised in a New York City apartment and had never had any contact with a bird before hearing this record.

The natural habitat of a wild animal is its Eden. We feel guilty when we move any particular wild animal to another place, even if the new conditions make its life easier and longer, probably because we, too, occasionally long for an animallike existence in Eden.

Our biological cradle, our Eden, was a semitropical savanna with few large trees, but a diverse vegetation and seasonal changes. Unlike wild animals, however, we human beings have spread all over the earth and most of us have now settled in environments to which we are not biologically adapted. For reasons that are not fully understood, representatives of *Homo erectus,* the immediate precursors of *Homo sapiens,* moved away from their biological Eden more than a million years ago and ever since the human condition has been increasingly different from animal life. Instead of living in nature, we modify natural environments in order to create artificial habitats that fit the biological attributes that we acquired during the Stone Age and that we retain wherever we settle on earth—and even when we move into outer space.

THE ORIGINS OF HUMANKIND

Several parts of the earth claim the honor of having been the cradle of the human species—but the decision will probably always remain uncertain because it depends upon what we precisely mean when we use the adjective "human." If "human" refers exclusively to anatomical and physiological characteristics similar to our own, then it is probable that the genus *Homo* originated several million years ago in the semitropical savanna of East Africa or, somewhat less likely, in a similar region of western Asia. The brain of these hypothetical precursors of humankind was much smaller than ours, approximately 600 cubic centimeters for *Homo abilis* as against 1000 to 1400 cubic centimeters for *Homo sapiens.*

The question of origin becomes less clear, and the answer consequently much more difficult, when the adjective human refers to the social, technological, behavioral, artistic and other cultural characteristics that we identify with contemporary men and women. Some of these characteristics were almost certainly present in *Homo erectus* which, or should we say who, in addition to standing erect, made simple tools and learned to use fire at least 500,000 years ago. *Homo erectus* seems to have been the first representative of the genus *Homo* to have moved away from Africa and thus to begin the human adventure which has led us to settle over the whole earth. *Homo erectus* occupied much of Europe and reached as far as Asia where we know him as Peking man, clearly identified with fire in the famous Choukoutien Cave. The size of *Homo erectus*'s brain was probably somewhat smaller than ours, but even this difference is questionable. It ranged from 730 to 1230 cubic centimeters and was therefore comparable to that of Anatole France, one of the most acclaimed twentieth-century writers—and a Nobel prize winner—whose brain measured only 1100 cubic centimeters! *Homo erectus*'s kit of tools was rather extensive, sufficiently diversified and sophisticated to enable him to live in parts of the earth to which he was not biologically adapted. He had therefore begun to set himself apart from nature.

Practically all the characteristics that we now consider distinctive to human beings can be recognized in people of Neanderthal and Cro-Magnon types who were living some 100,000 years ago. They were so similar to us and had so clearly become separated in many ways from the rest of creation that they deserve the name *Homo sapiens.* Their artifacts—from tools and weapons to sculptures and paintings—have great appeal for us not only esthetically, but also because they reveal preoccupations and activities that are important aspects of our lives, such as workmanship that goes beyond utilitarian necessity; tribal ceremonies; and burial of the dead, at times on beds of flowers. Equally remarkable are the precision and extent of the cognitive knowledge that these so-called "cavemen" had of the external world, including the awareness of its rhythms and natural laws.

We know too little about prehistory to agree on the exact time or place of this development; furthermore, there is some evidence that different groups of *Homo sapiens* achieved the human condition independently in several places on earth, more than 100,000 years ago. It has even been recently suggested that *Homo sapiens* first reached a high level of sophistication not in the Old World, but in California, from where humans moved to Asia over the Bering Straits.

One of the most puzzling and momentous events of human history has been the rapid replacement of Neanderthal people in western Europe by other people who looked more like us and who are usually called Cro-Magnon after the name of the cave in France where remnants of them were first discovered. The replacement of the Neanderthal by Cro-Magnon people occurred about 35,000 years ago, during a temporary thaw in the glacial age.

The Neanderthal people seem to have been the sole human inhabitants of Europe for at least 100,000 years. It used to be thought that they were brutish and primitive but in fact they walked just like us, erect and full footed. Their brain may even have exceeded ours slightly in size and they fashioned a distinctive kit of tools—those of the Mousterian culture. There seems to have been a high percentage of old people in their bands; the fact that every fifth individual among those who have been identified is over fifty years of age is amazing for any primitive hunting society. Two of the old Neanderthal people found in the Shanidar cave of Iraq were so severely crippled that they must have been completely dependent on the members of their group for a long time. Finally the Neanderthal people practiced ritualized burial of the dead, with flowers in the grave. Clearly they were truly human and for this reason are now referred to as *Homo sapiens neanderthalis.*

Most experts believe that the Cro-Magnon peoples migrated into Europe from Africa where their earliest fossils have been found. But this did not occur until approximately 35,000 years ago. By that time they knew how to manufacture complex tools, those of the Aurignacian culture. They also created wonderful artifacts such as the figurines known as "paleolithic Venuses" and the sublime and mysterious cave paintings of France and Spain. We now call them *Homo sapiens sapiens.*

Thirty-five thousand years ago, the two races of *Homo sapiens* finally met somewhere in Europe and the Neanderthal people rapidly vanished from sight, but the reason for their disappearance is still a mystery. It might perhaps have occurred through a rapid evolution of the Neanderthal into the Cro-Magnon type, an unlikely assumption; through a war between races, for which there is no evidence; through hybridization between the two races, a possible explanation; through failure of the Neanderthal people to adapt to environmental changes; or through some other process not yet recognized. These questions have inspired one of the great debates about human prehistory and since convincing evidence is not and may never be available, one of Europe's most learned evolutionary paleontologists, Bjorn Kurten of Helsinki, has elected to present all the relevant information, with a theory of his own, in the form of an exciting novel entitled *Dance of the Tiger.* I cannot evaluate the validity of Kurten's account of the disappearance of the Neanderthal people, but his novel has reinforced my belief that the various representatives of the species *Homo sapiens* soon acquired attributes and developed patterns of behavior that made them real human beings.

From Cro-Magnon time on, the evolutionary development of humankind has been almost exclusively sociocultural rather than biological. Humanity has transcended animality. The full human status had obviously been reached by the time the great axial religions were formulated more than 2500 years ago. However, some of our contemporaries, Joseph Wood Krutch for example, have expressed the view that while the biological species *Homo sapiens* has continued to prosper, humanity began to degenerate sometime during the late nineteenth century, when the wants of the consumer society took precedence over the cultural and spiritual aspirations of humankind.

There is at least one respect in which *Homo sapiens* has not significantly changed since the Stone Age. Whether they live in the temperate zone, in polar regions, in scorching deserts, or under humid tropical conditions all human beings still retain a genetic constitution best adapted to the savanna kind of country where our species acquired its distinctive biological characteristics millions of years ago. Wherever we settle, we establish a semitropical environment around our body, either with protective clothing or by heating or cooling our places of residence; if we develop settlements in densely forested areas, they are located in natural or artificial clearings; practically all the plants we use as food belong to sun-loving species. In fact, we are biologically so adapted to savannalike conditions that we could not long survive even in the temperate zone if we did not transform its natural environments so as to make them fit our biological needs and tastes. Whatever their skin color, place of birth or occupation, all human beings spend most of their time in the equivalent of zoos of their own making where they try to recreate the natural conditions of their biological cradle, a semitropical savanna. Each human group, however, modifies the natural environment in its own way, the result being an immense diversity of physical surroundings and of sociocultural characteristics.

We now live "out of nature," in the two very different meanings of this phrase. On the one hand, we live outside nature because our humanized environments have little in common with the natural ecosystems from which they were derived. On the other hand, everything we use comes ultimately out of nature even though much of it must be transformed before it can be used for human life. Paraphrasing Paul the Apostle, we are still *in* nature but no longer quite *of* nature. Since this profound change has been one of the most important steps in the humanization of *Homo sapiens* and in generating the human condition, the time and place of its occurrence might properly be considered to coincide with the real origin of humankind.

THE SOCIALIZATION OF HOMO SAPIENS

There is no convincing example of an animal learning to behave or to communicate in a genuine human way; not even the smartest chimpanzee comes close to it. In contrast, history provides countless examples of people from all parts of the earth who had never had any contact with Europe, yet acquired rapidly the behavioral patterns and the languages of the European people with whom they had to deal. Among the most famous cases are Malinche and Pocahontas, two young Amerindian women who played a crucial role in helping Europeans settle in the American continent.

Malinche was a young Aztec woman whom Hernando Cortez took as his slave shortly after his landing in Mexico. She became his mistress and also seems to have developed a strong emotional attachment to him. She learned Spanish and thus could serve as his interpreter with the Indians and also as an advisor on policies. She remained Cortez's inseparable companion and political ally even when the tide of war seemed to turn against him.

Pocahontas was the daughter of Powhatan, an aggressive Indian of the Virginia region who had created an Algonquian Powhatan confederacy consisting of some 9000 people in 1750. When the English tried to establish the second Jamestown colony, Pocahontas came into contact with them. She was then six to eight years of age and was much interested in the buildings and doings of those strange white people. Around the age of eleven she even seems to have fallen in love with one of their leaders, Captain John Smith, who was then twenty-five years old. Smith reported that she saved his life as he was about to be executed by the Indians. There is no doubt in any case that she brought food and other help to the members of the Jamestown colony who were at the point of starvation, and that she warned them of attacks by the Indians.

After being taken as a hostage on an English ship, where she was well treated and seems to have enjoyed ways of life new to her, Pocahontas eventually married an English merchant, John Rolfe, who took her to London on a business trip. In England she was treated as a real princess, even though she barely understood the meaning of the celebrations in which she participated. She became sick after a few months and her condition worsened when she was told by Rolfe that the time had come for them to return to Virginia, as she wanted to stay in London. In fact, she died on the Thames as their boat was departing and is buried near London. The legend is that she died "of a broken heart," but it is more probable that she had contracted tuberculosis or pneumonia, as has often been the case for semiprimitive people when they first came into contact with Europeans.

There are also many examples of persons from advanced societies who became parts of so-called primitive societies, in a few cases by choice but more often by accident or compulsion. As these cases are not well documented, I shall only mention that many Europeans who had had no prior contact with the wilderness rapidly adopted Indian-like ways of life when they moved to the underdeveloped parts of North America. The "voyageurs" or "coureurs des bois," for example, were among the most versatile and intrepid people of the American wilderness. In their search for furs they maneuvered their fragile canoes through explosive rapids and pushed their dog teams through subarctic blizzards. At the frequent portages they carried heavy loads, at times close to two hundred pounds, over rugged and often slippery rocks, almost at a trot. Most of these daring men came from the French Canadian villages along the Saint Lawrence River; they got on well with Indians whose ways of life they had to adopt and in many cases they married Indian women.

The Polynesians whom Captain Cook and le Sieur de Bougainville brought back to England and France from their explorations of the Pacific Islands in the eighteenth century became the rage of social life in London and Paris. Even the people of Tierra del Fuego, the crudest and most primitive population encountered by Darwin in the course of his voyage on the *Beagle,* did learn to speak English and adopted some European habits when they were taken to England. Harsh and coarse as life was in Tierra del Fuego, it took place in a structured human society that enabled Fuegians to assimilate other human cultures and languages.

Children who have been deprived of human contacts show the extent to which early social conditioning is essential to make *Homo sapiens* capable of acquiring the patterns of language, behavior and culture that so obviously differentiate human life from animal life.

Despite the common use of the phrase "wolf children," the caring for children by wolves has really never been proven. There may be some validity to the phrase nevertheless in view of recent reports that wolves can feed human children by regurgitating food as they do for their own pups; furthermore, feral children tend to walk on all fours and can follow the pack. There are several well-documented cases of boys and girls who have lived in the wild with little if any human contact until their adolescent years. Such children were usually in good physical condition when first discovered and brought back to a human environment, but their behavior was so unlike that of children of their age group raised in association with other human beings that they could be said to be socially naked.

The best-studied feral child is the "wild boy of Aveyron," who was twelve to thirteen years old when first seen emerging from a forest in the mountains of central France during the extremely severe winter of 1799. He could then run rapidly on all fours, climb trees, conceal himself amidst the vegetation, stand bitter cold and nourish himself with wild plants and raw potatoes that he took from the fields. After being apprehended, he managed to escape several times thanks to his stamina and physical skills.

When finally confined to a house, he swayed back and forth like a monkey in a zoo, behaving in a disgustingly filthy manner, tearing away any article of clothing put on him, scratching and biting the persons who tried to establish contact with him and to feed him. On the other hand, he displayed great excitement with bursts of laughter and almost convulsive joy when nature put on a spectacular show, for example, when the sun was shining and the wind blew from the south. On beautiful nights, when there was a full moon, he would awaken and remain motionless, in a contemplative mood of ecstasy, interrupted by deep sighs and faint plaintive sounds.

A young physician, Dr. Jean Marc-Gaspard Itard, became interested in the boy and adopted him under the name of Victor. He undertook to socialize and educate him, watching and recording every aspect of his behavior and particularly every sign of socialization. In a few years, Dr. Itard taught Victor to understand some French and to express a few desires in a simple language, but he never succeeded in making him really engage in conversation. He managed to have him accept some clothing but could not instill in him behavioral patterns that would have made it possible for him to live in a normal French society. Although Victor became somewhat more sociable with time and even showed affection for the woman who took care of him, as well as for Dr. Itard, he repeatedly tried to escape and could not be made to behave in a way compatible with social life. He eventually had to be placed in an institution where he died of some infectious disease at thirty-three years of age. Dr. Itard was convinced that Victor was fundamentally normal and that he acted like an idiot only because he had been abandoned in the woods in his very early childhood.

Several other studies of feral children, as well as of children grossly deprived of human attention, have confirmed that the effects of isolation during early life are always disastrous and may be irreversible. Of particular significance is the case of an American girl who is referred to as Genie, both to protect her identity, and to convey "the fact that she emerged into human society past childhood, having existed previously as something other than fully human." Details concerning her family and history have been reported in detail in *Genie,* a book by Susan Curtiss, then at the University of California in Los Angeles, who has devoted herself to the socialization of Genie after she was discovered in 1970 as an adolescent who had been deprived and isolated to an unprecedented degree.

Genie is the product of a very unhappy marriage. Her father disliked all children. A boy born before Genie was subjected by the father to such rigid rules of obedience and discipline that he manifested early developmental problems. He was late to walk and talk, had eating difficulties and was still not toilet trained at the age of three. At that point, his paternal grandmother took him into her own home; he eventually became a normal child and was returned to his parents.

When Genie was seen by a pediatrician at the age of five months, she was noted to be alert and of normal weight. She was seen again when eleven months old, was then slightly underweight but was described as alert, capable of sitting alone and with normal primary dentition for her age. At sixteen months of age she developed an acute pneumonitis and was seen by another pediatrician who found her feverish, listless and unresponsive. Genie's father, who was intensely jealous of the attention her mother paid to her, used the pediatrician's statement as justification for her subsequent isolation and abuse.

She was confined to a small bedroom, harnessed to an infant's potty seat, unclad except for the harness. Unable to move anything except her fingers, hands, feet and toes, she was left to sit, tied up most of the time for several years. At night, she might be removed from her harness only to be placed into another restraining garment, a sleeping bag which her father had fashioned to hold her arms stationary. She was then put into an infant's crib with wire mesh sides and a wire mesh cover over her head.

When Genie was thirteen and a half years old, her mother, though blind, succeeded in getting in touch with her own parents, who took her and Genie to their home where they stayed for three weeks. A health worker became aware of the situation and Genie was hospitalized for extreme malnutrition in November 1970. The police were alerted and the father committed suicide on the day of the trial.

Having hardly ever worn clothing, Genie did not seem to suffer from either heat or cold, an imperviousness to temperature that had been reported for the Aveyron boy, Victor, and for other "wolf girls." In the hospital, Genie was completely silent even in the face of frenzied emotion. Videotapes of her behavior during her first months of hospital life reveal that although she understood a few words she "could not process a sentence of English on the basis of its linguistic content alone, but rather that she depended critically on gestures and other nonlinguistic cues to make any sense of speech directed to her." Her performance on various tests to measure her cognitive abilities placed her around the two-year-old level.

She was moved in December 1970 to a rehabilitation center which offered greater opportunities for socialization, a richer activity program and better access to the outdoors than did the hospital ward. Her physical state improved rapidly and by April 1971 tests for cognitive abilities placed her at the four to six years level. When Susan Curtiss, the author of the book concerning Genie, began to work with her, in June 1971, Genie could understand and use a few words, but her behavior was unpalatable. The book presents many examples of disgusting personal habits and of other behavioral peculiarities that were completely unacceptable socially.

In June 1971, she was moved to a foster home—a warm loving household with two teenage boys, an adolescent girl, a dog and a cat. For almost two years, she talked primarily in one-word utterances, then in two-word strings for the next two years, but progressively she learned to express herself more fully and to achieve some control over her feelings and behavior. Although she still could not read at the time the book was written in 1976, Susan Curtiss states that "Genie continues to change, becoming a fuller person, realizing more of her human potential. By the time this work is read, she may have developed far beyond what is described here."

Thus, belonging to the species *Homo sapiens* is not sufficient to provide all the attributes that make us fully human. We learn to become human by hearing human speech and observing human behavior during the critical years of childhood. All members of the species *Homo sapiens* have in common certain fundamental biological and mental characteristics which constitute their nature, but this nature can take a human form and generate human ways of life only when exposed to suitable conditions—the nurture—for its development.

THE INVARIANTS OF HUMANKIND

We become only one of the many persons we are capable of becoming. All of us are born with a wide range of potentialities that enable us in theory to develop an immense diversity of attributes, but in practice we develop only those aspects of our nature that are compatible not only with the conditions to which we are exposed, but even more with the choices we make in the course of our lives. The marvel is that nature and nurture can become so completely integrated that they generate a unique socialized entity, the human person, out of the biological organism *Homo sapiens.*

Each person is unprecedented, unrepeatable and unique. Not even homozygotic—so-called identical—twins are identical in real life. They have the same genetic constitution, but nevertheless become different persons because they are exposed to different environmental conditions, first *in utero* and even more so after their birth. On the other hand, while human beings come to differ profoundly from each other as they live in different ecological niches, they can all interbreed and thus remain members of the same species *Homo sapiens.* Different from each other as we are, we all have in common many fundamental characteristics and needs that can be called the biological and behavioral invariants of humankind and that play essential roles in all the sociocultural expressions of human life. These invariants are found throughout our species regardless of economic, social, ethnic or national status.

One of the greatest scientific achievements of recent decades has been the demonstration that the hereditary biological characteristics of all living organisms, which are their invariants, are transmitted by DNA molecules which constitute genes. I had the good fortune to witness the first phases of this discovery which occurred in the early 1940s at the Rockefeller Institute for Medical Research in the microbiological laboratory where I was then working.

In early 1944, my colleagues Avery, McLeod and McCarthy published in the *Journal of Experimental Medicine* a paper in which they demonstrated that they could modify at will one of the hereditary characteristics of the microbe that causes lobar pneumonia—the pneumococcus—by growing this microbe in a culture medium containing the deoxyribonucleic acid (DNA) obtained from another pneumococcus which was naturally or artificially endowed with this particular characteristic.

In all living organisms—from the smallest to the largest, animals or plants, human beings or bacteria—the DNA molecules of the genes are the carriers of hereditary characteristics. All DNA molecules have the same fundamental chemical structure, but minor differences among them and in their arrangement along the chromosomes are sufficient to account for the phenomenal diversity of living species and for the peculiarities of each individual organism in a given species. On the one hand, the general pattern of the DNA molecules and of their arrangement determines whether the creature will be a hare, a horse, or a human being. On the other hand, subtle differences in the DNA patterns of each species determine the hereditary characteristics of each individual hare, horse or human being.

The fundamental potentialities and needs that I have referred to as the invariants of human nature can be satisfied in so many different ways that they may be difficult to recognize in the usual manifestations of human life. A few examples may therefore be useful to illustrate how the fundamental uniformity of *Homo sapiens*—the invariants of human nature—can express itself in the prodigious diversity of human life under the influence of nurture—the environmental and sociocultural conditions.

The nutritional habits of vegetarians appear at first sight diametrically opposed to those of carnivorous people. Many African people and a very large percentage of Asians feed almost exclusively on vegetables, tubers and fruits. In contrast, the East Africa Masai people are nourished almost exclusively from what they derive from their herds of cattle, including the blood they draw from these animals every day. Despite these profound differences in nutritional *habits,* however, all human beings, whether vegetarian or carnivorous, have essentially the same *requirements* with regard to intake of calories, carbohydrates, fats, amino acids, minerals, vitamins and other essential chemical nutrients. Diverse as they may appear, all human diets provide similar mixtures of these chemical nutrients, granted that the total intakes differ according to age and the ways of life.

From the nutritional point of view, it makes little difference whether the essential nutrients are derived from plants or animals, or even from products obtained by chemical synthesis; they can provide adequate nourishment if consumed in adequate amounts and proportions, provided of course that they are not contaminated with dangerous microbes or chemicals. The required mixture of nutrients is thus an invariant of human nature whereas the kinds of foodstuffs we eat are its sociocultural expressions. That these can take many different forms is illustrated by the fact that cooking recipes are among the most distinctive characteristics of national, regional, social and cultural groups.

Another invariant of human nature is that all normal human beings need enclosed shelters or at least protected areas into which they can retreat either for safety and comfort or simply to withdraw from public contact. Stone Age people commonly had access to caves and they built simple huts; countless kinds of shelters have been used or built throughout the ages. On the other hand, human beings enjoy open vistas and probably even have a psychological need for them. In the not too distant past, a common way to punish a child for misconduct was to have him or her face a wall for a certain length of time as this was known to be an unpleasant experience. The essential visual needs of humankind can be met in different ways—by clearing areas for settlements in the temperate or tropical rainforest, by lawns in front of homes, by extensive views from the top of a hill, a mountain or a skyscraper.

Whether we satisfy our nutritional requirements with a vegetable dish or a steak, our need for shelter by retreating into a natural cave or behind the closed shutters of a cozy room, our longing for open space by contemplating a seascape, an undulating field of grain or a classical parterre, we deal with fundamental and universal invariants of human nature in as many different cultural ways. And so it is for other invariants. For example:

The brains of all humans are genetically endowed at birth with special structures located in the so-called Broca area that make it possible to learn *any* of the existing thousands of human languages. Some persons have mastered more than twenty languages, but the immense majority of us learn only the language of the particular social group in which we are born and raised.

As has always been the case in the past, young people crave adventure and sexual satisfactions, adults are eager for achievement, most old persons long for quiet and stability; but these universal urges are governed by countless codes of behavior peculiar to each ethnic group, each society, and each period. In the course of time, the games of love have been played in the countless ways which have been represented in as many forms of literature and art; the eagerness for achievement may find its expression in political power, the accumulation of wealth, or the discovery of a natural law; quiet and stability may be found in the management of one's garden, in a daily visit to a park or within a community.

While the fundamental basis of our behavior is still what it was millennia ago, its social expressions are culturally determined and have changed in the course of history. Homer's heroes still interest us because we are moved by passions similar to those that motivated them; but the gods and adventures of the Homeric tales have now been replaced by powerful public men engaged in political or economic conflicts.

From time immemorial, human life has derived its color from dancing, music, poetry, fiction, painting, sculpture, tattooing, body painting and festivities that transcend obvious biological needs; but these expressions and celebrations have greatly differed from time to time and from one social group to another. Pageantry has involved situations as different as those symbolized by the Lascaux paintings, the Stonehenge circle, the Buddhist temples, the Greek agoras, the Gothic cathedrals, the Renaissance palaces, the Victorian ceremonies, the Arches of Triumph everywhere and the ticker tape parades on Broadway—all manifestations that would have different meanings or no meaning whatever for members of the species *Homo sapiens* who had been raised in a different culture. It is by incorporating an immense diversity of sociocultural patterns into biological life that *Homo sapiens* becomes truly human.

Biological uniformity is readily explained by assuming that all members of the species *Homo sapiens* have the same origin, and that they have continued to interbreed despite the differences generated by life in their various ecological and sociocultural niches. Social diversity, in contrast, has multiple causes, most of which are poorly defined. It derives from minor genetic differences between groups and persons, from the influence exerted on development by environmental forces, from the uniqueness of individual experiences, from the artifacts and institutions created by each society, from traditions, imaginings and aspirations—all of which are the consequences of innumerable choices.

In addition to the influences that reach us from the outside—the world external to us—there are other influences that exist in the individual mind of each one of us and that constitute our private conceptual environment. Whether primitive and poorly informed, or sophisticated and learned, each one of us lives as it were in a private world of his own. In fact, the conceptual environment may be more influential than the external environment because it affects all aspects of our lives—

the ways we deal with ordinary experiences, our views of man's place in the order of things, how we conceive of natural laws and even the attributes we associate with the word God. Our direct contacts with reality may have less importance for the shaping of our person and of our life than our individual and collective dreams.

The social contrasts between Athens and Sparta, between the Vikings and the troubadours, between the Zuni and the Apache obviously depend on factors more numerous and complex than racial characteristics, or than the climate, topography and geology of the regions where these people developed and lived. Neither do economic patterns account for the cultural differences among Europeans. City-states and nation-states emerged, not from natural forces but from historical and social occurrences that created various conceptual environments in the different populations.

Environmental influences are always complicated and often completely distorted by our universal tendency to symbolize everything that we experience; we respond to these symbolic distortions as if they were reality. In most cases we do not create these symbols ourselves; we receive them from the social atmosphere in which we live. A gray sky on November 1 is simply depressing and boring for a person raised in New York City but it creates in me a poetical mood because it evokes the gentle sadness of All Saints' Day in Paris. The first of May has violent political connotations for many Europeans but is for others the one day of the year when couples explore the woods under the pretext of gathering lily of the valley. Eating a special dish may cause nausea or help the secretion of digestive juices depending upon the circumstances under which it was first eaten.

Our views of the physical and social universe are impressed upon us by rituals and myths, taboos and parental influences, traditions and education—all mechanisms which provide us with the basic premises according to which we conceptualize our inner and outer worlds. The process of socialization through which *Homo sapiens* becomes really human consists precisely in the acquisition of the collective symbols characteristic of one's social group, with all their associated values. Admittedly, most symbolic systems change with time. Observance of the sabbath and of dietary laws is not as strict among the reformed as among the orthodox Jews; divorce and eating meat on Friday are no longer as sinful as they used to be for people of the Roman Catholic faith. In general, however, symbolic systems persist for many generations within a given culture, even though they may change form. Concepts about the universe and behavioral patterns are thus transmitted as a social heritage which minimizes individual differences within a given group, or at least masks them, and thereby gives greater homogeneity to the group.

Practically all factors responsible for human diversity are interrelated, but I shall nevertheless discuss them separately for reasons of convenience. I realize, however, that any analytical separation of these factors results in a false image of human life. Despite its diversity, humankind consists in the manifestations, endlessly ramified, of the various aspects taken by *Homo sapiens* under the combined influences of the cosmic, biologic, and cultural forces that have generated the seemingly endless spectrum of human societies.

2

THE PAST, PUBLIC PLACES, AND SELF-DISCOVERY

LIFE AS EXPERIENCE

SURVIVAL OF THE PAST

NATURAL AND BUILT ENVIRONMENTS

IMAGES OF HUMANKIND

CHOICES AND CREATIVITY

SELF-DISCOVERY

2

THE PAST, PUBLIC PLACES, AND SELF-DISCOVERY

LIFE AS EXPERIENCE

Whether plant or animal, small or large, no living organism can function as an independent, separate entity. To live implies not only utilizing available resources but also being shaped by them, modifying them and thereby achieving a state of intimate integration with the total environment. Living organisms can be understood only when they are considered as part of the system within which they function. This is particularly true of us human beings because all aspects of our lives are profoundly influenced by an immense diversity of physical and cultural forces which shape our bodies, our behaviors and the social structures to which we must relate in order to become fully human. Mnemosyne, who was the Greek goddess of memory, was also a symbol of Life and the mother of the nine muses who are the moving spirits of creativeness. Her complex nature in the Greek myth symbolizes that our individual lives always involve the creation of novelty; we progressively become what we are at any given moment because we can use both conscious and unconscious memory to incorporate the past into present conditions.

I cannot think of myself as a person without calling to mind countless surroundings and events which I remember with some precision; and without realizing that many influences of which I was not aware at the time of their occurrence have also left on me permanent imprints.

The common use of the phrase *living substance* reveals how impoverished is our perception of the richness and subtleties of life. There is no living substance. Whether we deal with microbes, melons, mice, or human beings, these creatures cannot be dealt with merely as substances or objects as long as they are living. At any level, life implies the integration of an immense diversity of substances which, functioning as a unit, continuously interplay with their particular environments, often in creative ways. For the amoeba as well as for the elephant, to live is to experience and function.

The greater the freedom of a particular organism to select where it goes, what it does, and how it responds to stimuli, the more complex and more creative is the living experience. Individual representatives of a given type of wild mushroom, morels for example, are much the same wherever they grow; even though gypsy moths can move, they also are just typical representatives of their species. In contrast, a particular dog or cat will change profoundly in appearance and behavior if it escapes from the house where it has been treated as a pet and elects to live in the woods. Human beings have the greatest degree of freedom and therefore the greatest range of creative adaptability.

There does not seem to be any way to demonstrate scientifically that we are endowed with freedom. In fact, philosophical reasons make it likely that it is not possible for the human brain to achieve a complete understanding of its own working and that the existence of free will must therefore be accepted on faith, as an expression of the living experience. In any case, lack of scientific proof does not weigh much against the obvious manifestations of free will in human life and perhaps in other forms of life. As Samuel Johnson wrote two centuries ago, "All science is against freedom of the will; all common sense for it." Biological sciences have developed a great deal since Samuel Johnson's time, but not enough to justify the claim by orthodox behaviorists that purely deterministic mechanisms completely account for all types of behavior. As noted by two Nobel laureates in two famous biological laboratories, one at Harvard University, the other one in Belgium, "Under the most perfect laboratory conditions and the most carefully planned and controlled experimental procedures, animals will do what they damned please. . . . Could one ask more free will than that?" These two biologists did not question that all phenomena of life are conditioned by heredity, past experiences and environmental factors, but they acknowledged by this statement that certain animals, and human beings much more so, can choose among several possible courses of action, thus transcending the constraints of biological determinism through a faculty which is conveniently referred to as free will.

In practice, free will simply means that persons, and probably many animals also, but to different extents, can visualize alternatives to any situation and choose one of them. We differ from animals by our much greater ability to imagine future situations in a distant place and to choose on the basis of such imaginings. I love both New York and Paris, can afford to live comfortably in either one of these cities, and have precise knowledge of their respective advantages and disadvantages. Not a week goes by that my wife, born and raised in the United States, and I, born and raised in France, debate as to where we should settle for the rest of our lives. My wife leans toward Paris for reasons that are not entirely clear to me whereas I prefer Manhattan for reasons that are no better than hers. In any case, both of us have freedom to move and we shall continue to exercise this freedom on the basis of selection among imagined alternatives.

I shall therefore take free will for granted, simply because I believe that human beings constantly make choices and take decisions that give the lie to absolute biological and behavioral determinism, but I shall nevertheless first consider certain aspects of human life in which the person involved cannot control either the environment or its effects, and therefore has little if any chance of manifesting freedom of response or action.

SURVIVAL OF THE PAST

We try to be rational in most of our activities, and find it reassuring that we can recognize and control many of the influences that affect us, even when we do not understand the mechanisms or the effects of these influences. However, many aspects of our lives are largely beyond our control because they are the consequences of events that occurred in the distant past and of which we are not even conscious.

The extraordinary degree to which our physiological processes are linked to cosmic rhythms provides a striking illustration of the persistence of traits that emerged millions of years ago during the evolutionary development of the human species. We tend to believe, for example, that we have achieved independence from the forces of nature because we can illuminate our rooms at night, heat them during the winter, and cool them during the summer; also because we can secure an ample and varied supply of food throughout the year. But even when we function in an environment which seems constant to us because we can control several of its elements, all the functions of our bodies continue to fluctuate according to certain rhythms linked to the movements of the earth, of the moon, of the sun and perhaps also of other parts of the cosmos. While we can control heat, humidity, light, food intake and a few other components of the place where we live, our bodily mechanisms exhibit daily, seasonal, and perhaps other rhythms which certainly affect our physical and mental well-being.

Our biological and psychological responses to any stimulus are different in the morning from what they are at night, and different in the spring and summer from what they are in autumn and winter. There was a sound biological basis to the Indians' practice of attacking white men just before dawn, because the physiological and psychological defense mechanisms of human beings are then at a low ebb. The wild imaginings of the night and the fears which they engender are indirectly affected by the earth's movements, in part at least because of diurnal and seasonal changes in body levels of the various hormones. It is common experience that the operations of the human organism are more likely to escape from the control of reason under the influence of darkness and particularly at certain hours of the night. And it is also a well-established fact that the effect of a given toxic substance, or of a medicinal drug, can greatly differ according to the hour of the day and according to the season.

Lunar cycles are also probably reflected in our physiology and behavior. There is evidence that moon worshippers as well as "lunatics" are really affected—as the words suggest—by lunar forces to which most of us probably respond in some way. In apes and probably in human beings also certain physiological processes related to sexuality seem to be heightened during periods of full moon.

Seasonal changes affect us so profoundly even if temperature and illumination are artificially maintained at a constant level that their influence is reflected in social practices. Many of these first emerged in primitive social groups and have continued in different forms even as societies have become more sophisticated. Today in the most mechanized, treeless, and birdless urban environments, just as in the legendary Arcadias of long ago, men and women perceive through their senses and reveal by their behavior the exuberance of springtime and the despondency of late fall.

Seasonal patterns of behavior may have their origin in the fact that organic processes as fundamental as the secretion of hormones and the way food is metabolized in the body differ from season to season even when environmental conditions are artificially controlled so as to make them seemingly uniform throughout the year. Although these physiological phenomena are of practical importance, they are poorly understood and in fact have been hardly studied. It has long been known, for example, that in rats kept without food for forty-eight hours the levels of so-called acetone bodies (ketosis) are approximately three times as high from May to October as they are during the winter months; winter levels remain low even when the animals are in a room maintained at summer temperature. The high level of summer ketosis is associated with a low capacity of the rats' tissues to metabolize glucose—a deficiency probably due to a reduced functional activity of the pancreas. It is probable that in human beings also, some of the seasonal metabolic patterns have their basis in variation of hormonal activity. Blood pressure, urinary excretion of nitrogenous compounds, deep body temperature are among the many physiological activities that have been shown to exhibit seasonal variations.

Biological reasons more complex and more subtle than mere changes in temperature certainly affect seasonal patterns of behavior, for example the European custom that carnival and Mardi Gras are celebrated when the sap starts running up the trees, and that the dead are commemorated on November 1—All Saints' Day—when nature appears to be dying. Many legends and ceremonies of ancient people that appear to be purely cultural traditions in fact had a biological origin in the seasonal relationships of human beings to their environment. Much of the Greek mythology associated with Demeter, Persephone and Adonis, as well as the corn dances among American Indians, can readily be interpreted as early local practices related to seasonal conditions.

Even when we are in heated or air-conditioned dwellings we, modern people, continue to be under the influence of cosmic forces much as if we lived naked in direct contact with nature. We also continue to react to the presence of human rivals or of certain animal species as if we were in danger of being attacked by them. All over the world, a large percentage of people display a deep horror at the mere sight of snakes and spiders at an early age, with little more than gentle nudging from their parents. In contrast, children who are constantly warned not to go near electric sockets and automobiles, or not to play with knives, rarely develop phobias against such objects. A possible explanation for these behavioral differences is that, in the distant past, human beings had unpleasant or dangerous encounters with snakes and spiders and that these experiences became in some way inscribed in the biological memory of our species.

The so-called "fight and flight" response, with all its deep physiological accompaniments, is almost certainly a biological carryover from the time when encountering a wild animal or a human stranger made it a matter of life or death to mobilize the body mechanisms that permitted our distant ancestors to engage in physical struggle or to flee.

Many other natural situations of the distant past are still reflected in our responses to present social situations. For example, we still experience physiological disturbances when lost in the wilderness, not only in a tropical forest or a desert, but also in an unfamiliar urban agglomeration among people who commonly generate in us a sense of panic if we are not familiar with their ways.

More generally, the effects of crowding, of isolation or of unexpected challenges have effects that reflect responses to analogous situations in the evolutionary past. However, responses that were once favorable for biological success may no longer be suitable under modern conditions. For example, suspicion of the stranger was biologically useful in the Stone Age but now usually takes dangerous forms such as racism or xenophobia. Phenomena ranging from the aberrations of mob psychology to disturbances of metabolism and of blood circulation which result from arguments at the office or at a cocktail party may be to a large extent the survivals of attributes which were useful when they first appeared during the evolutionary development but which are irrational, useless and possibly dangerous under the conditions of modern life.

Changes in blood pressure, in the distribution of blood within the various parts of the body, in the secretion of hormones such as those secreted by the adrenal and thyroid glands, and more rapid utilization of blood sugar are among the physiological responses that helped animals and early people to fight or run away. One can almost take it for granted that, under similar conditions, changes also occurred in the secretion of the newly discovered brain hormones, the endorphins, that can alter our perception of pain. In the modern world, any threatening situation mobilizes these and other mechanisms even though the threats rarely lead to physical conflicts or efforts. It has been observed for example that the coach of a rowing crew watching the performance of his team from the shore undergoes physiological changes similar to those of the athletes actually engaged in the contest.

The urge to control property and to dominate one's peers is also an ancient biological trait which can be recognized in the different forms of territoriality and dominance. Even the play instinct corresponds to an important biological need which exists in animals and has probably always been part of human nature because it helps the infant to discover the world and to learn functioning in different situations.

These and many other biological characteristics are woven in the very fabric of the human race, and they condition all aspects of human behavior. Most of them are probably encoded in the genetic endowment of the human species although some may be transferred culturally, from generation to generation. In any case, it is difficult or perhaps even impossible to investigate innate responses by the orthodox analytic methods of science which are based on detailed studies of the component parts of the organism. Like free will and the mind, most biological clocks and responses of the total person to complex social situations disappear when the studies are carried out on organs or cells separated from living organisms. The most interesting phenomena and experiences of life can be observed only when the organism responds to its total environment as a whole integrated unit. Fortunately, the investigations of life in outer space and in submarines are creating a healthy wave of interest in the biological problems concerning the organism as a whole—such as the effects of the tides, of the seasons, of the diurnal, lunar and annual cycles—effects which have been grossly neglected so far by biomedical sciences. Just as rockets and satellites are giving new importance to celestial mechanics, so do the prospects of prolonged sojourns in space or under water call attention to the need for better knowledge of the cosmic forces that have shaped life on earth and that still influence us today.

When watching the sky on a clear, cloudless night, even we jaded modern people perceive that we are part of a seemingly boundless universe extending beyond the stars, and that we participate in its rhythms. Ancient people probably experienced this feeling of oneness with the cosmos more intensely than we do. Even chimpanzees have been reported sitting absolutely still watching a sunset, as if fascinated by the spectacle. Ancient people were aware of a regularity in the movements of the celestial bodies as far back as the Old Stone Age; they recorded the phases of the moon by scratching notches on objects of bone or ivory and recorded the seasonal movements of animals and the growth of plants. The observation of astronomical events must have played very early an important role in the mental life of ancient people if it is true, as seems to be the case, that immense megalithic monuments such as the Stonehenge circle in England, the Carnac alignments in France, the great Giza pyramid in Egypt were oriented in such a manner as to provide spectacular views of sunrise at critical times of the year. It has been claimed that in Egypt, more than three thousand years ago, the birth of the sun was celebrated at midnight on December 25, when the sun was in the sign of Capricorn, also known as the stable, and that the sun, in its course from the winter solstice, was in the sign of the Lamb at Easter.

Philosophers, writers, and artists have always been aware of the role played by occult processes in human life. In Plato's dialogue *Phaedrus,* Socrates speaks of the creative forces released by "mania," the "divine madness." The text of the dialogue makes it clear that the word "madness" as used by Socrates refers not to pathological mental states but rather to those deep biological attributes of human nature which are almost beyond the control of reason and usually not even recognized, except by their effects on behavior. These attributes may remain concealed under the usual circumstances of ordinary life, but they constitute powerful sources of inspiration for the artist and for the scientist as well. Creativity usually requires hard work, but it depends even more on intuition and inspiration. Hearing "the voice of the deep" enables us to tap resources from regions of human nature which have not yet been thoroughly explored.

Nietzsche was referring to innate forces analogous to Socrates' divine madness when he wrote in *The Birth of Tragedy* that the Dionysian inspiration is a necessary complement of the Apollonian attitude which considers reason and order as the highest values. As shown by E. R. Dodds in his book, *The Greeks and the Irrational,* ancient civilizations were aware of the existence in human nature of powerful biological urges which were not readily controlled by reason. The occult passions—Socrates' divine madness—were commonly symbolized by a ferocious bull struggling against reason.

All over the world, social practices have emerged empirically to allow that these occult forces manifest themselves under somewhat controlled conditions. The Dionysian celebrations, the Eleusinian mysteries and many other rituals served as release mechanisms for biological urges which could not find an otherwise acceptable expression in the usual ways of Greek life. As rational a person as Socrates participated in the Corybantic rites with their music and ecstatic dancing. Ancient practices related to the seasons still persist in the most advanced countries of the Western world, though often in a distorted form, as can be seen in the outlandish costumes and behaviors during the carnival celebrations in many parts of the world. The paleolithic bull survives in the urbane city dweller whose own way of pawing the earth becomes manifest whenever a threatening gesture is made on the social scene or when seasonal changes activate the various hormonal mechanisms.

We thus relate not only to our physical, biological and social surroundings but also to the cosmos as a whole even when we are not conscious of this relationship. Whatever the type of social organization and however primitive human beings may seem to us, all human societies have developed myths and rites to express their partnership in the cosmic system by activities that transcend their purely biological needs.

The response to the ringing of bells, for example, is not so much to the sound waves themselves as to their symbolic overtones. Endlessly modified as sound spreads in all directions through space, the pealing bells symbolize for me that we are related to everything in the cosmos. They reach into the great beyond where they lap on the shore of the ultimate mystery which science may never be able to solve.

I used to think that my strong emotional response to church bells was merely a consequence of my having been conditioned to their religious significance by my Christian upbringing in France, but this is not the whole explanation. Hearing the muezzin at close range in the Arab section of Jerusalem evoked in me feelings very similar to the ones I experience when listening to the bells of a Christian church. And the same applies to other sounds associated with rites that I have witnessed in many parts of the world—the gongs in the Buddhist temples of Japan and Taiwan, the chants at a Navajo tribal ceremony, the rhythmic voices during a Polynesian dance in Tahiti, the drums of black people in central Africa. Even the most confirmed atheist can experience unity with the cosmos from the ringing of bells, the reverberations of a sonorous gong or the hypnotic effect of chants in New Mexico or drums in Africa.

The cosmic experience derived from bells or other ritual sounds contributed to the development of animism among primitive people. Animism persists as an undercurrent in all great religions which, in their highest form, involve the response of the person as a whole to the universe as a whole. In some way or another, we may soon have to accommodate our relation to the cosmos with the theory that everything in it began abruptly at a definite moment in time, some twenty billion years ago, with a tremendous explosion, a flash of light and energy—the Big Bang. Discussing the emotional response of theoretical physicists to the Big Bang theory of creation, the astronomer Robert Jastrow, himself an atheist, recently stated that these responses "come from the heart whereas you would expect the judgments to come from the brain." As scientists reach the point of knowledge beyond which they cannot go, they are still, in Jastrow's words, "greeted by a band of theologians who have been sitting there for centuries."

The theologians, in fact, were not the only ones to be waiting and watching. Perceptive and sensitive human beings have always wondered about the origin and destiny of the cosmos and their place in the order of things. The ultimate questions have always been those inscribed at the bottom of one of Gauguin's Tahitian paintings: "Where do I come from?" "Who am I?" "Where am I going?" The church bells announcing a Christian celebration and the minaret's muezzin summoning the Moslems to prayer symbolize that, whatever our system of belief, we perceive human life as more than the chemical reactions that assure the maintenance of its anatomical structures and physiological functions. To be human implies not only being part of a society but also attempting to relate our existence to creation as a whole.

The DNA molecules of our genetic endowment determine the invariants of our biological nature. These include behavioral patterns that were developed by our precursors during the Old Stone Age or by our more recent progenitors and that govern our responses to environmental and social stimuli. The forms and intensity of these responses at any given time are profoundly influenced, however, by our own individual experiences. We can never escape from our past individual conditioning, especially that arising from the experiences of early life. In a split second, a fragrance sends us plummeting to the deepest layers of our being; even the faintest odor can thus reconnect us to another place and another time. Bad smells associated with the places where we have become conscious of ourselves are probably as powerful as the pleasant ones in evoking the feeling of security and the joie de vivre associated with many days of our youth.

I remember the intense pleasure of a young girl returning to the squalor of her native Appalachian coal town after her first experience away from home in an idyllic rural setting. She took a deep breath of the murky, sulfurous air and radiantly exclaimed—"Oh! home!" I understand her reaction because I also experienced a similar Proustian pleasure when I perceived, after half a century of absence, the stench of a sugar beet distillery in the French village where I had been raised.

In one of his letters home while at school, the young Louis Pasteur expressed intense longing for a whiff of his father's tannery even though the old methods of tanning created smells that were nauseating. The spoken word also reflects the indelible stamp of early influences. I am told that I write and speak English correctly and at times elegantly, but both my pronunciation and certain turns of phrase betray that I was twenty-five years old before I shifted from French and Italian to the English language.

I am still so much a part of the world in which I was raised and it is still so much a part of me that thinking of the Île de France landscape is sufficient to transport my mind to the villages where I see myself as a little boy wearing a smock and pushing a wheelbarrow full of tall grasses and wild carrots gathered in the fields for the rabbits at home. I still feel under my feet the softness of the country lanes; I smell the hawthorn in the hedgerows; I hear the mooing of the cows and the songs of the birds, especially of the larks arising from the wheat fields; I expect any time to catch sight of the church steeple in the center of the village.

While the little boy with the wheelbarrow survives in me, I am of course now very different from him, and not only because I am old. I have lived in many different places and have associated with many different persons, either by accident or design. Few are the parts of Paris, Rome, London, and mid-Manhattan which do not instantly play back in my mind entire situations in which I participated and that made me forever different from what I was before their occurrence. I do not live in the past; it is the past which is alive in me.

NATURAL AND BUILT ENVIRONMENTS

The various human races exhibit obvious physical differences which have emerged as a result of exposure for many generations to different surroundings and ways of life. Some of these differences are hereditary because they have been encoded in the DNA molecules which determine the genetic constitution peculiar to each human race. This is the case for the skin pigmentation characteristic of racial groups, for the small size of the African pigmies and the Australian aborigines, and probably for a few minor anatomic differences between Japanese and

Caucasians. Some students of child behavior claim that babies of the different human races exhibit minor differences in developmental rate, posture and temperament which are genetically determined. Gesell tests have recently revealed, for example, that the motor development of a certain group of African infants was greatly in advance of that of European infants of the same age and was paralleled by advance in adaptiveness, language acquisition and personal-social behavior. The precocity of African infants was usually lost in the third year, although it was retained by some of those who had the advantage of a kindergarten education.

However, most of the physical and behavioral characteristics that are distinctive of ethnic groups are in reality not genetic but the consequences of differences in sociocultural conditions. For example, most immigrants from Sicily were of short stature when they landed in New York at the turn of the century, but their children and especially their grandchildren born and raised in the United States are now likely to be as tall as the descendants of the original settlers from northern Europe. Similarly, the Jewish people who lived in central European ghettos before the war had a multiplicity of distinctive characteristics that were assumed to be the expressions of "semitic" genes, but their children born and raised in the Israeli Kibbutzim are tall, lanky and tend to depart widely in their behavior from the traditional patterns that prevailed among the Jews of central Europe before the war. Many postwar Japanese also are much taller than their prewar parents and grandparents. Human development is thus profoundly affected by environmental forces and ways of life which act so rapidly—within one or a very few generations—that their effect cannot possibly be due to genetic changes.

The shaping of certain human attributes by the natural environment had been clearly recognized more than 2000 years ago by Chinese and Greek physicians, although it is unlikely that these scholars clearly differentiated between genetically inherited characteristics and those individually acquired. The Greek physician Hippocrates was one of those who emphasized that the physical and mental characteristics of the various populations of Europe and Asia, as well as their military prowess, were determined by the topography of the regions in which they lived and especially by the local quality of air, water and food. Hippocrates taught for example that "inhabitants of rocky, mountainous, well watered countries . . . tend to have large built bodies adapted for endurance and courage" whereas inhabitants of low, warm, swampy lands tend to be thickset, fleshy and indolent. We now understand that the differences observed by Hippocrates were the expression not of genetic differences but of the relative prevalence of certain infectious diseases and nutritional deficiencies in the places where he made his observations.

During the eighteenth century, the Abbé Jean Baptiste DuBos (I am not related to him) was one of the most articulate French exponents of the doctrine that we are largely shaped by geographic and especially climatic factors; he was in this sense an intellectual descendant of Hippocrates and a predecessor of Montesquieu. DuBos emphasized the effects of weather not only on human development, but also on the emergence and manifestations of intellectual attributes. His words, "Le climat est plus puissant que le sang et l'origine," can be translated as meaning that climate has more powerful effects on the body and the mind than have the constitution of the person and the country of origin. He believed that the quality of the air influences the composition of the blood and thereby all physical and mental characteristics. According to him, the effect of the climate explained why the uncouth and vigorous Frank and Norman barons who had settled in Mediterranean countries had become "effeminate, treacherous and pusillanimous," and why the Arabs had lost much of their stamina after settling in southern Spain. While the factual observations were historically correct, DuBos' explanation is almost certainly erroneous. The Norsemen and the Moslems had become weaker in the Mediterranean environment not because they had been genetically affected by the climate but because luxury and leisure had prevented them from cultivating the physical stamina and mental virtues which had accounted for their military triumphs.

Scientifically primitive as they were, the views expressed by the Abbé DuBos are nevertheless valuable because they conveyed the important truth that surroundings and ways of life have profound effects on many aspects of human character and development, but his suggestion that climate has a profound effect on intellectual ability has unfortunately been used to support racist doctrines.

Early in our century, the geographer Ellsworth Huntington of Yale University strongly upheld the doctrine of climatic determinism in the several editions of his very successful book *Civilization and Climate.* He held the view that climate influences not only food production and human health but also intelligence as well as morality, and derived a racist theory from these assumptions. In his words, "The climate of many countries seems to be one of the great reasons why idleness, dishonesty, immorality, stupidity, and weakness of will prevail." Believing that a temperate climate is more conducive to progress than a tropical one, he went as far as to conclude that "the Negro seems to differ from the white man not only in feature and complexion but in the workings of the mind." As he saw it, the climatic effects experienced by black people during their biological evolution in tropical Africa had made them inferior human beings. In passing, it would be interesting to know how he would have reacted to the recent finding, mentioned above, of remarkable precocity among certain African infants.

In a parochial expression of superiority, Huntington even claimed that eastern New England, the very region where he lived, has a level of seasonal variability just sufficient to provide a degree of challenge which is stimulating but not overpowering, and is therefore ideal for human development and for civilization.

The view that proper levels of variability stimulate human performance and development is of course a special aspect of Toynbee's "challenge and response" theory of civilization. But while Huntington's emphasis on the importance of climate is based on a few valid physiological observations, it constitutes a very incomplete and thereby fundamentally inaccurate statement of the effects of natural forces on human character and development. In particular, it fails to explain why some of the major civilizations of the past have emerged and developed in natural environments where seasonal variability and other natural conditions are very different from those prevailing in eastern New England or Europe—for example the Egyptian civilization in the Nile Valley between immense stretches of desert, the Inca civilization at the high altitude of the Peruvian Andes, the Mayan and Khmer civilizations among the massive forests of the humid tropics. History shows that human beings have been capable of prodigious cultural achievements under a great variety of natural conditions and it shows also that periods of intellectual doldrums have occurred repeatedly in regions that Huntington regarded as ideal for human life and for civilization.

One of the reasons human development can be successful in many different types of natural environments is that *Homo sapiens* is still essentially a semitropical animal, so that throughout history and even prehistory most human beings have been *biologically* out of place in the regions of the earth where they have settled. In order to colonize the earth, we have had to create almost everywhere, out of the wilderness, *artificial* humanized habitats that enable us to function and to multiply in *natural* environments to which we are not biologically adapted. We could not survive long even in the so-called temperate zone if it were not for the *social* practices we have developed to protect ourselves against the inclemencies of the weather, and against food shortages during the winter months.

Under normal conditions, we spend most of our time and carry out most of our activities not in the wilderness, not even in open nature but in highly humanized landscapes, urban agglomerations, towns, villages, and rooms of various sizes and shapes. These are the environments which exert the most profound influences on our physical and mental nature. For lack of precise knowledge, these influences can be stated only in the form of hypotheses.

In general, the pace of life is less rapid in the country than in large cities, as documented by actual measurements. Surveys have shown that people in European villages or small towns amble at a speed of approximately three feet per second whereas the average rate of walking is five to six feet per second in cities of more than a million inhabitants—whether in European cities such as Prague (Czechoslovakia) or American cities such as Brooklyn (New York). In mid-Manhattan, the rate of walking is distinctly faster on Park Avenue where the shop displays are rather subdued than on the side streets where there is much more to inspect.

This difference in pace does not necessarily mean that life is less strenuous in the country than in the city. The more general truth is that almost all aspects of the built environments in which we live markedly influence the way we function and therefore probably the way we develop physically and mentally.

The craving for adventure, which is widespread and powerful among young people, corresponds primarily to a search for new experiences. Although the exploratory drive and the need for certain minimum levels of stimulation exist in all primates and probably throughout the animal kingdom, these characteristics seem to reach their highest levels in the human species. In the past, the hunger of youth for the unexpected could often be satisfied simply by exploring around the home—whether on the family farm with its highly diversified activities, or in the city where streets used to differ from each other and where each block had many different facades. In modern urban agglomerations, environmental uniformity decreases the opportunity for such innocent adventures—a fact which suggests that many restless, energetic pioneers might have become juvenile delinquents if they had been compelled to live in our kinds of towns and cities.

Attitudes are also influenced by the design of dwellings and of furniture. One does not behave the same way in the austere simplicity of a Japanese house, in the disciplined elegance of a classical European living room, or in the sloppy comfort of modern overstuffed furniture. Tatami-covered floors and fragile paper walls give the Japanese house unique acoustical properties that may have influenced the design of musical instruments and even the pitch and timbre of Japanese voices. The sound of a piano loses much of its brilliance in traditional Japanese houses, but on the other hand a Japanese samisen loses much of its subtle quality in the reverberating box characteristic of most European and American buildings.

A recent newspaper item will illustrate how human behavior and development can be influenced by the design of homes.

In August 1977, the local newspaper of a New York City suburb published the plan of a house designed to emphasize energy conservation. One component of the architectural concept was a combination eat-in kitchen and family living room with a fireplace, so arranged that this subunit could be essentially sealed off from the rest of the house during the cold months. The architect assumed that the family would spend many hours within this "retreat," enjoying togetherness by the warmth of the fireplace, and could thus save on the amount of energy required to heat the entire house. This kind of house design, however, did not appeal to one of the readers of the article. "Too much togetherness is for the birds," he wrote to the newspaper. "Wrecking family tranquillity is much too high a price for saving a bit of energy." According to him, the most successful feature of modern American houses is precisely to have made possible the separation of the different familial activities. His ideal is that the various members of the family have their own separate lives "without getting into each other's hair" by using living room, family room, dining room, recreation room, separately whenever wanted, according to their individual fancy.

A few years ago, a point of view exactly opposite to that of the newspaper reader just mentioned had been expressed by Dr. A. L. Parr, former director of the New York Museum of Natural History, who was born and grew up in a small Norwegian town at the turn of the century. The rooms of the house in which Dr. Parr was raised were individually heated by fireplaces or stoves. Light was provided at first by candles, then by lamps using kerosene, acetylene or gas. As such techniques of heating and lighting were somewhat cumbersome and dangerous, they could not be entrusted to children and created some difficulty even for aged people. In consequence, the whole family—children, adults, and oldsters—had to spend most of their evenings together, enjoying or at least tolerating each other's company. The phrase "family circle" may have originated from such groupings of children and parents around the source of light or near the hearth. My own experience in a French village at the beginning of this century, and in Paris during the First World War, has been much the same as that of Dr. Parr. I agree with him that the necessity to function for several hours in close contact with a diversified group of people made it necessary to cultivate social tolerance, or at least a discipline that facilitated other types of social relationships.

The introduction of central heating and of electric lights has decreased the necessity for family togetherness. These new techniques, being safe and convenient, can be safely operated even by children who can thus have freedom of behavior in their own living quarters. Such a change in the ways of life has obvious advantages but may also have undesirable consequences. A room of one's own gives a sense of freedom and may help in the development of individuality, but the behavioral outcome may be a loss of social discipline and cohesion. Having to study, read and play in contact with adults for several hours around the family lamp imposed restraints on one's behavior, but constituted a training for the inescapable difficulties of later social contacts.

Human behavior and development is also certainly influenced by the general planning of human settlements.

The visual appearance of the streets in the French villages where I spent my early youth is gray and banal, indeed visually unattractive. Most Americans see them, as I am often inclined to do now, as a dull monotonous wall which calls to mind an institution in which people are out of contact with public life. This interpretation of the appearance of many French streets has some validity. The people I knew during my childhood rarely entered houses other than their own, not even their neighbors' houses, except by special invitation. In the two villages and the small town where I spent my early years, I do not recall having been inside more than a very few houses in each, even though my parents and grandparents were on good terms with the people they knew.

However, while the walls lining up the French streets commonly look grey and uninviting, the homes behind them usually had an intimate atmosphere best described by the word *foyer*—hearth. Ever since prehistoric times, the hearth has been the physical and spiritual center of the fundamental social unit—whether extended or nuclear family. The houses behind the uninviting wall were in my youth the places where the families held a life so private that the emotional links it created could never be destroyed, not even by cultural or familial alienation. The words of a French peasant that I recently read sound quite familiar to me and rather appealing. "Comme nos pères, nous sommes méfiants. Il fait froid dehors et chaud dedans. . . . On dit bonjour sur le pas de la porte, on ne fait pas entrer n'importe qui. . . . La vie se déroule à l'intérieur entre la pièce principale et l'étable. . . . La grand-mère écarte le rideau, apprécie le temps. . . . Le grand-père remet une bûche dans la cuisinière." "Like our fathers, we are suspicious. It feels cold outside and warm inside. . . . One greets people as they pass by in front of the door but few are the ones we let in. . . . Our real life is spent within the home, between the main room and the stable. . . . The grandmother opens a corner of the curtain to get an idea of the weather. . . . The grandfather then puts another log on the woodstove." Each language has its own phrase for "at home," "chez soi," "a casa"—each expressing a different nuance of familial relationship but all denoting an invariant need conveyed in French by the word "appartenance" and in English by the word "belonging."

In all places I knew during my French youth, there existed behind the nondescript wall and adjacent to the house a good-sized garden, not only for flowers, but also for vegetables, chickens and rabbits to be used in home cooking. The translation of the word garden can be treacherous. The French word *jardin* has come to evoke for many people the gardens of the classical Italian and French landscape designers. In contrast, the English word garden is chiefly associated in my mind with what I have known in the United States. To my taste, the American garden is almost objectionably public. I cannot imagine any more glaring contrast as a background for social relationships than the enclosed intimate foyers and gardens behind the dull but protective walls of French streets, and the picture windows and lawns opened to the public eye which are almost the rule in American settlements.

The French people I knew were not as antisocial as the design of their communities suggested. Their social life was carried out in public places, whether cafés, restaurants, shops, public parks or gardens. On the other hand, the Americans even now are not as socially opened as would appear from their unwalled residences, lawns and gardens; their social gatherings are visible to everyone but nevertheless segregated in many different ways, some subtle but nevertheless effective.

The academic person knows that he does not quite belong when invited to speak before a group of automobile salesmen or before the Bel-Air social group in Los Angeles. A Christian may not feel quite at ease in a Jewish gathering and vice versa. As to the color of the skin, it is officially ignored, but I do remember a most friendly discussion in a social meeting with a highly educated black man who assured me that it was impossible for me, as a white person, to really understand the feelings of the black community.

The type of French community in which I grew up led to a range of social acquaintances that was rather limited but to associations that were usually deep and long-lasting. This was true for hostile as well as for friendly relationships. The American type of settlements result in a much wider range of human contacts for which a caricature was provided to me by the wife of an army officer who had been stationed for less than two years in a small Connecticut city. As she and her husband were about to leave that city, she stated her dismay at the thought of losing the more than five hundred new friends they had made in less than two years. She would no doubt soon make five hundred other friends in the place of her husband's new assignment. I often wonder whether my early conditioning in a very restricted part of France, the *Vexin français* and the *pays de Thelle,* accounts for the fact that, despite offers of interesting and more lucrative jobs in other parts of the United States, I have spent practically all my professional life on the Rockefeller campus in New York City and have lived most of the time within twenty minutes walk of my office.

Behaviors and tastes naturally change. The walls on the main street of the village where I was born, Saint Brice-sous-Forêt, still look as grey and uninviting as they were at the time of my birth. On the other side of the railroad track, however, a similar village, Sarcelles, became the site of large housing developments after the war. Although the new housing facilities in Sarcelles were better than those in the old village, the changes of habits they required increased the incidence of a variety of diseases among the inhabitants; the pattern of diseases was so ill defined that physicians referred to it as "sarcellitis." Now, three decades later, the inhabitants of Sarcelles have become adjusted to their new environment and furthermore have modified it to make it conform better to the traditional ways of French life. Sarcellitis disappeared spontaneously once this had been achieved. My native village, St. Brice, also has recently grown, but in a more subdued way. Its new houses are small, and although the lawns and gardens surrounding them are enclosed, the fences are low and give a somewhat open character to the landscape. It is not, however, the open freedom of the American landscape, only an openness compatible with the still persisting French desire to enclose family life within a frame, even though it be only a symbolic frame.

Historical accidents certainly played a large part in shaping human settlements in Europe and in the United States. Of special importance was the abundance of land in the New World which made it easier to achieve relative independence by building farmhouses widely separated from each other.

Invariants of human nature also played a part in the design of human settlements. Life in the savanna of our biological origin, and over most of the earth during the Ice Age, created dual and opposite kinds of visual needs in human beings. From the opening of the Cro-Magnon cave in Les Eyzies the eye ranges over a wide horizon where Stone Age people could follow the movements of the game, while the cave constituted a shelter into which they could retreat for protection against dangerous animals or inclement weather. Ever since prehistoric times, styles and practices of planning have attempted to satisfy these two complementary needs of our species—the need for the protective coziness of a shelter and also for ease of visual contact with the outer world.

In addition to the physical and social surroundings, other aspects of the environment profoundly alter human growth and the quality of life. For example, the nutritional habits, the kinds of diseases most prevalent in a given area at a given time, and the stimuli that pervade a particular community are biological factors that profoundly influence physical and psychological characteristics.

The maturation of young people has greatly accelerated during the past few decades in all countries that have adopted the ways of Western civilization. Not only are children taller than they were a few decades ago; final adult heights and weights are also greater and are achieved earlier in life. A century ago, maximum stature was not reached in general until age twenty-nine, whereas it is now reached at about nineteen in boys and seventeen in girls. Sexual maturation is also advanced. Whereas the mean age of menarche was around seventeen in 1850, it is now around twelve in affluent countries.

The factors responsible for the dramatic changes now being observed in the rate of physical and sexual maturation are not completely understood. Improvements in nutrition and control of infections—both for the mother and the child—have certainly played a large part in the acceleration of development during early childhood; this change in turn probably contributes to the larger size of adults. There is some evidence also that greater ease of communication results in a wider range of choices for a mate and that the consequent increased matings between various human groups expresses itself in what biologists call hybrid vigor.

Although little is known of the long-range behavioral consequences of changes in the rate of biological development, it can be assumed that earlier anatomical and physiological maturation exerts an influence on the ease of finding one's place in the social order of things. It may also affect certain psychological attitudes and even the forms of civilization. As the Japanese grow heavier and taller, for example, they are likely to change the design of their furniture, buildings and grounds—even the management of their landscapes and the conduct of their ceremonial practices. The tea ceremony may not be compatible with the attitudes and gestures of husky teenagers who spend much of their days clad in blue jeans.

In his essay "On the Uses of Great Men" R. W. Emerson remarked that "there are vices and follies incident to all populations and ages. *Men resemble their contemporaries even more than their progenitors* [italics mine]." The corresponding Arab proverb, "Men resemble their own times even more than they do their fathers," similarly expresses the general truth that most of our biological and behavioral traits are profoundly affected by surroundings and events. We resemble our progenitors because we derive our genetic constitution from them, but we resemble our contemporaries to the extent that we experience the same environmental conditions as they do and are therefore conditioned by the social and physical peculiarities of our times.

To a large extent, the diversity of built environments reflects the changes of tastes in the various human societies. As a teenager in France, my first visual experience of modern architecture was a photograph of the Woolworth Building, then the highest and most famous landmark of New York City. Immediately upon its completion in 1913, this skyscraper was known all over the world as an architectural wonder—not only a bold feat of engineering but even more a clarion call for a new kind of architecture suited to the technological age. Gothic frills ornamented the Woolworth Building from top to bottom to signify that it was the modern version of the medievel cathedral, but a cathedral built to celebrate the power of money rather than the glory of God.

The name "Cathedral of Commerce" which became attached to the Woolworth Building symbolized that this type of architecture had become the expression of a mercantile society.

The Woolworth Building remained the highest skyscraper until 1929. But despite its sixty stories, it looks small now in comparison with the 107 stories of the twin towers of the World Trade Center recently erected not far from it. Furthermore, the complexity of its external design and decorative features, harkening back to outmoded religious and social allusions, make it appear old-fashioned to our eyes which have been conditioned by the austerity of the modern international "functional" style. The impeccable, elegant creations of Mies van der Rohe in Chicago, New York and other American cities, the Trade Center in New York, the even higher Sears Tower in Chicago, the startlingly chic and arrogant Hancock Building in Boston, the incredibly bold Pennzoil-Zapata Building in Houston, were not designed to be "cathedrals" of modern life but as the workplaces and dwellings of an unromantic society, striving for its own kind of sophistication.

In the late nineteenth and early twentieth century, the pioneers of the international functional style expressed their architectural theories in a few arresting formulae. The phrase "Form should follow function," attributed to the American architect Louis Henry Sullivan, was meant to convey the view that the physical appearance of a building should reveal the function of each of its parts and should not be encumbered with unnecessary adornment. Mies van der Rohe's "Less is More" implied the belief, so admirably expressed in several of his own buildings, that the very simplicity of design contributes to its esthetic quality. When Le Corbusier referred to apartment houses and private dwellings as "machines à habiter," he meant that architectural design should aim primarily at the efficient performance of housing function, as is the case for the specialized functions of other machines of modern technology.

All these statements, rational as they may sound, make real sense only if the word "function" can be clearly defined. There has been a tendency among modern architects to restrict the relationship between form and function to the structural aspects of design. According to this view, the form of a building should acknowledge its supportive parts and its mechanisms for the various services but the most important functions of a building are its human uses. Functionalism should therefore take into account primarily the physiological and psychological well-being of the occupants, and esthetic satisfaction as well as symbolic significance for the persons who experience the building from the outside. Practicality and comfort have little importance in the functionalism of monumental churches, palaces, fortresses, prisons, arches of triumph, tombs of important personages; what matters in such buildings is their psychological significance. Functionalism should naturally refer also to certain social needs and to ecological impact.

Since it is and will always be difficult to provide successfully for all the functions served by any particular building, form usually follows that particular function that the designer or the user has elected to emphasize. The complaints of the public against modern architecture come largely from the fact that many architects of the functional school seem to have been more interested in the technical aspects of function that in its human or social aspects.

Some of the functions to be served by architectural design correspond to the various moods of the human mind: sublimity, beauty, coziness, sadness, fear, awe, respect, admiration, etc. Each period and each part of the world has its own historical or regional way of expressing these different moods. The emphasis on massiveness in Egyptian temples and pyramids conveys a religious attitude and a kind of relation to nature radically different from those conveyed by the striving for height and light in Gothic cathedrals. Modern architecture has often produced anonymous machines for just living or working, as if human beings had no other preoccupations than these, and as if the architects were satisfied with manufacturing disposable cubicles for dispensable people. Furthermore, there is now a widespread feeling that modern architecture, despite its attempts at visual novelty and its purported rationality, has become as formalistic in its own way as was the Beaux Arts architecture it began displacing almost half a century ago, and that it does not do justice to the most important human and social functions of our times.

Architects are accused of identifying with engineers and of no longer thinking or feeling as artists. Their buildings may be functional from the structural point of view, but not from the human point of view. Modern buildings are said to provide physiologically uncomfortable living and working conditions; to create a sterile and depressing atmosphere for the people who function in them and for those who look at them; to disregard the need for the intimate warm relationships associated with gregariousness and affection; to insult eye, ear and touch; to spoil the quality of their physical surroundings. Last, but not least, modern buildings are commonly wasteful of energy and space. The present retreat from the modern style constitutes an attempt to reintroduce human and ecological factors into the architectural equation.

Some recent textbooks of architecture devote many pages to the effects of environmental factors on physiological processes. The authors consider this approach essential because classical builders and planners have been primarily concerned with visual effects, as if other sensual experiences were not as important. In fact, pictorial representations of a building or landscape can be misleading even though visually accurate. The best painting, photograph, rendering or model captures only certain visual appearances under a limited set of conditions but fails to convey how a building sounds, how it smells, how it feels. The total sensuous impact is one of the true measures of architectural achievement. Whether in the Parthenon, in Epidaurus or in a Cape Cod cottage, what counts in the end is the total experience. Design should take into consideration all aspects of human physiology and ecology.

At its best, modern architecture involves not only the incorporation of sophisticated technologies into the construction and use of buildings, but perhaps even more attempts to achieve a rational esthetic quality. There are many aspects of human life, however, that cannot be analyzed in rational terms. While steel-framed and glass-walled buildings may provide accommodations more comfortable and practical than do traditional buildings, they commonly generate a feeling of alienation because they do not speak to the subrational needs which are probably as important as the conventional values. We like buildings and feel well in them, not only because they are rational in design and construction, but even more because their total atmosphere and symbolic significance fit our emotional longings. It is difficult if not impossible to verbalize these sensual and perceptual values of architecture, let alone to quantify them. For this reason, they will remain in the domain of the architect as an artist rather than becoming part of his science. Their role in planning can be illustrated by an example.

Places that are somewhat mysterious and rooms that are unusual in shape have an appeal that seems irrational, yet is nevertheless real. Children in particular, but adults also, commonly prefer attics, cellars, and other removed areas in which to carry out their hobbies; the desire for mysterious places may have a biological basis in the human instincts for exploration and withdrawal. In any case, the irregularities of old human settlements, such as the Mediterranean hill villages and towns, are magnets for visitors and tourists who live in buildings that are physically comfortable and efficient but where nothing is unexpected and mysterious. Pascal's "le coeur a ses raisons que la raison ne connait pas" may apply to the fact that rationality in building and planning does not satisfy some of our deepest emotional needs.

In our times, the most important contribution to the philosophy of design has been the recognition that the external forms we give to our environments reflect some aspects of our inner psychological states. Societies create great architectures only to the extent that they value certain ways of life. The present coldness and ugliness of many modern cities is the concrete expression of our social ills. Some commercial and industrial buildings are architecturally more imaginative than those we build for social relationships or religious worship, simply because we are giving more emphasis to the materialistic aspects of life than to its spiritual aspects.

There may be reason for hope, however, in our increasing awareness that social and cultural considerations are values to be emphasized in the creation of desirable environments. The conscious manipulation of our built environments—of the human landscape—may eventually be a factor in the consensus that will create the humanism of the Machine Age.

IMAGES OF HUMANKIND

Human development thus involves much more than the biological processes through which the infant is progressively transformed into an adult and old person. Becoming human implies the passage of *Homo sapiens* out of nature into culture. For each person, this passage results from an evolution which is guided, and indeed to a large extent imposed, by the set of assumptions of the social group to which that particular person belongs—assumptions which influence practically every aspect of individual life.

The systems of beliefs and the patterns of behavior of a given society create in the social group a coherent image of humankind which integrates concepts as diverse as the origin and purpose of life; the relationships between human beings, animals, plants, land, water and sky; the rules of conduct and practically all other aspects of man-in-society and man-in-the-world. The various sets of assumptions thus result in as many different images of humankind and of its place in the order of things. These images are inevitably reflected in landscapes, buildings and social structures but, paradoxical as it may sound, they do not seem to be much influenced by the natural environments. In fact, very different types of society based on different images of humankind can coexist under the same natural conditions.

In the American Southwest, for example, the Pueblo people developed long ago a highly structured communal way of life in which order and moderation were the ruling virtues and unanimity was expected in all important social decisions. Acceptance of such a strict set of rules resulted in a peaceful society but did not leave much room for originality in individual development. In contrast, the Spanish society that settled in the American Southwest during the seventeenth century prized highly the cultivation of individualism but did not allow behavior to conflict with the authority of either the church or the state. Later immigrants into the Southwest, coming from all parts of Europe, organized their societies around much more individualistic concepts. Among them, each person was expected to create a place of his own within society, in fierce competition with other human beings and also with nature. Every man for himself was the formula for self-creation and material success. Thus, three very different images of humankind became established side by side in one small part of the American continent.

Under usual circumstances social conditioning occurs chiefly through the upbringing of children—how they are held and sung to; what they are fed and the stories they are told; whether they are expected to stay close to home or allowed to wander just for the sake of adventure; and most importantly what they are taught concerning the present state of affairs and the expected shape of things to come. Children constantly exposed to an atmosphere of rising expectations and to pictures of a fanciful electronic world cannot avoid having a conditioning different from that of children raised in traditional cultures which convey patterns of behavior through stories of an ideal past and through fables in which the behavior of animals symbolizes virtues and vices. Since my experience of children's upbringing is extremely limited, I shall restrict myself to a report of a few facts of my own childhood experiences and try to recognize how they have shaped my images of humankind and my subsequent life.

I have pleasant memories of my very early childhood but most of them are rather vague. An important aspect of it was that I developed acute and extremely painful rheumatic fever around age seven, with extensive cardiac damage which has persisted and which still prevents or limits certain types of physical activity which would otherwise appeal to me. This early physical handicap has certainly affected my subsequent behavior and has perhaps compelled me to derive most of my satisfactions from emotional experiences and intellectual pursuits. I might not be spending time on this book if my heart enabled me to play tennis or to go jogging.

I spent my early years in small farming villages up to age thirteen and was never far from horses and from the various animals that were kept on our grounds before being slaughtered in a building adjacent to my father's butcher shop. For me, animals were clearly expendable creatures. Watching animals and helping to feed them was part of my daily life and this experience may explain why I do not recall ever having cuddled a teddy bear, ridden a broom-handle horse, or needed any other substitute for real animals. There was a big dog called Capitaine in our household but although I played with it I never had the depth of attachment for it which commonly develops between pet and child. I still like playing with dogs and cats for a short time but I have never had a real pet and find it more entertaining to watch animals in the wild. Some of my earliest and most endearing memories are of our home vegetable garden and of its simple but fragrant flowers. I have to refrain from writing here the French names of the flowers most common in the Île de France and which evoke for me the seasons of my youth. My early emotional contacts were thus with humanized nature rather than with toys or fanciful objects as substitutes for reality.

I do not recall ever being told fairy stories. My father and mother were so busy running the shop that they could not have found time for storytelling. The only two of my grandparents with whom I had frequent contact were very earthy people, not of the fairy tale type. Participating in house and garden chores, digging worms to go fishing from the banks of the Oise River, sitting on the sidewalk in front of the house after supper, and carefully closing the shutters at night are the most vivid memories of the days I spent in my grandparents' house.

My sister and brother are younger than I, and my associations with them were friendly, but did not play an important role in my life. I see myself on a float around the age of nine dressed up as a little marquis next to my slightly younger sister in a similar outfit, along with other children in a special village celebration. The memory of this event has convinced me that there is great value in the communal activities of a neighborhood but I have enjoyed them much more as an observer than as a participant. My wife and I are always "somewhere else" on Christmas or New Year's Day, and if possible, at the time of familial or office parties. I started school at age five, and almost every aspect of my life related to my school years is as vivid as if I were living it now.

I learned to read and write quickly and was especially good in history and geography, except for the fact that I was never able to draw a decent map—as was then expected of children in a French school. I had to read fairy tales, but without much interest; they were probably too unrealistic in comparison with the earthiness of my daily life. Paradoxically, however, I was enthralled by stories of medieval knights, their ladies and their daring adventures. It seems odd that stories of *Le Petit Chaperon Rouge* and *La Belle au Bois Dormant* lacked appeal for me because they were unrealistic, whereas I read avidly the most outlandish tales of courtly or martial adventures in medieval life. The heroic and romantic Middle Ages provided the dream world in which I lived most intensely until I discovered the American Far West in weekly accounts of "Les Aventures de Buffalo Bill" in French magazines. Somewhat later the excitement of American city life reached me through tales, also in French magazines, about the detectives Nick Carter and Nat Pinkerton.

Schooling introduced me very early to another world of the mind much closer to reality than the Middle Ages or America. The village where I spent most of my youth had only two one-room schoolhouses, one for the boys and the other for the girls—worlds apart, even though on the same small village square—each with forty to fifty children. As we moved up from one grade to the next, we were given the responsibility of helping to teach the lower classes. This turned out to have been for me an excellent educational system, in part because it compelled me to review constantly what little knowledge I had, but more importantly because it helped to instill into me a sense of responsibility and some ability to relate to other children in an orderly way.

Another important aspect of my education was that, from the day I could count and be trusted (around eight years old) I helped my mother run the butcher shop during the busy hours of the day; I sat at the primitive cash register, collected the money, and handed back the change. I have long believed that the most useful part of my education, from the human point of view, was to watch my mother converting the sale of a lamb chop into a pleasant social event.

In school, teaching naturally followed the rigid French program of those days but two special aspects of it deserve to be emphasized because I now realize that they are still influencing my life today. Every morning the teacher dictated a classical text, of increasing difficulty as we moved up in age. Without leaving my village, I thus progressively discovered a world of places and people very different from the one in which I lived. It was from dictations that I learned through Flaubert the life of a servant in a provincial town, through Tolstoy the behavior of peasants harvesting wheat in Russia, through Chateaubriand the experience of spending a night in an awesome forest of the New World.

Every day we also had to learn by heart and recite a piece of classical literature. La Fontaine's fables occupied an important place in this assignment and I remember many of them. Each of these fables conveyed a message intended to make us, children, wiser about the ways of the world. I doubt that we were much interested either in the stories or in the messages, yet the fables may have influenced our attitudes more than we realized. Now, seventy years after learning them by heart, I still quote to myself in French some of La Fontaine's stories or sayings that fit some particular situation in which I am involved. Since this may sound like rationalizing for the sake of romanticizing my youth, I shall mention here three fables that directly relate to my present life, and that I do tell myself in French while working.

Every spring for more than thirty-five years, I have planted trees with my own hands on an abandoned farm I own in the Hudson Highlands, fifty miles north of New York City. As I plant the trees, there always comes to my lips the beginning of La Fontaine's fable, "Le Vieillard et les trois jeunes hommes" (The Old Man and Three Youths). The story is about three young men who jeer while watching an old man setting out trees, telling him that he should not work so laboriously since he will not live long enough to benefit from this labor. To which the old man gently replies:

Thinking of fruit can give us twofold felicity
Here, prospectively, then in the days Fate has in store.

Eventually, each of the three young men dies by some accident and the old man mourns them by carving the stories of their lives on their gravestones. I continue planting my own trees, telling myself, as did La Fontaine's old man, that someone will enjoy the shade they will cast after I am gone. I shall continue planting trees as long as I can even though this must seem to those who watch me the act of "a dotard, no doubt" as the three young men said in La Fontaine's fable.

Whenever I hear that the world is about to run short of natural resources, I recite to myself the first lines of La Fontaine's fable, "Le

Laboureur et ses enfants" (The Husbandman and His Sons). The fable is about an old farmer who, realizing that he is about to die, calls his sons to his deathbed to tell them that his own parents buried a treasure somewhere on his land. Although he does not know where the treasure is buried, he assures his sons that:

If each of you will search for it with hardihood
You are sure to find it; you will come on it at last.

The sons follow their father's advice, do not find any treasure, but work the land so thoroughly that it produces larger and larger crops, thus fulfilling the father's final words:

If you'd find a fortune, work hard.

For our time, however, the most important part of the fable may be its second line:

Travaillez, prenez de la peine
C'est le fond qui manque le moins

which means, according to me, that we have to revise our views about natural resources. Even farmland has to be created, and its fertility maintained by human effort. As I shall discuss in Chapter 5, what we call "natural" resources are, in reality, the raw materials of the earth which have to be transformed into useful products by knowledge, imagination and hard work.

I could speak of many other of La Fontaine's fables which frequently come to my mind in the daily affairs of my existence, but I shall limit myself to one more, namely, "Pierrette et le pot au lait" (The Dairymaid and the Milk Pot). Pierrette, a dairymaid, is shown walking to the market with her milk pot on her head, imagining along the way all that she will be able to do by investing the money that she will get from the milk; first, she will buy eggs, then progressively raise chickens, pigs, and cows. But in her excitement, she falls; having spilt her milk she leaves us with the lesson that we should not count our chickens before they are hatched. I find it often useful to remember this lesson.

Most of La Fontaine's fables that I can still recite deal with human beings, but there are many more fables where animals symbolize human attitudes. The role of animal stories, like that of animal toys, goes much beyond mere amusement. It creates mental pictures which prepare the child for the real world by feeding the imagination and forming moral and ethical templates. This dual role was well recognized by La Fontaine himself as can be read in statements taken from his own preface to the fables. According to him, a fable is "an imaginary episode used as an illustration; all the more penetrating and effective because familiar and usual. Anyone who offered us but master minds to imitate would be affording us an excuse for falling short; there is no such excuse when ants and bees are capable of performing the tasks we are set. . . . Nor are the fables just a good influence; they also extend our knowledge of the modes of behavior of animals and thus of ourselves, since we epitomize both the good and the bad in creatures of restricted understanding. . . . So the fables are a panorama in which we see ourselves."

In his classical study of *La Fontaine et ses Fables,* the nineteenth-century French philosopher Hippolyte Taine tried to explain the characteristics of the fabulist by exploring those of the Île de France region surrounding Paris, the landscape in which La Fontaine spent most of his time. Comparing this region with northern Europe, the German forests and areas of wild mountains, Taine found in the Île de France a host of sensations "which explain what it is to be French." A trip that he took to the North Sea, Holland and Germany inspired him to write as he returned to La Fontaine's country: "The landscape is not majestic or powerful; the atmosphere is not wild or sad; monotony and moody spirit disappear; variety and gaiety begin. There is neither too much plain nor mountain; neither too much sun or humidity. Everything . . . is on a small scale, in convenient proportions, with an agreeable and delicate air. The mountains have become hills, the forests woodlands. . . . Small rivers run among trees with a gracious smile. . . . Here are the beauties of our landscape; it appears rather flat to eyes accustomed to the noble architecture of the southern mountains or to the heroic and abundant vegetation of the north, but its grace stimulates without exalting or overpowering the mind."

If it is true, as Taine claims, that La Fontaine's character and poetry had been shaped by the general features of the Île de France region, these must have influenced also the symbolism attached to the stories told in the fables. Since the fundamental aspects of human nature are the same everywhere, fables all over the world probably try to teach similar lessons, but with regional differences both in the nature of the messages and in their forms. Although I have recognized some of these differences while reading fables from different countries I doubt that we can ever really understand the subtleties of another culture, especially the symbolism of the message it conveys to children through fables. Having lived in Italy in my early twenties, at an age when I was highly receptive and readily adaptable to any new situation, I nevertheless found it difficult to appreciate the full meaning of certain Italian attitudes even when I could imitate them successfully, and even though French and Italian life have much in common. I still smile inwardly today when I go to my Italian barber in New York while he is eating a sandwich in his shop and welcomes me with a warm "Vuol favorire?" as if he really wished me to share his meal. And being referred to as "illustrissimo" still disturbs me somewhat today even in my most conceited hours. In fact, there are many aspects of American behavior which are still somewhat baffling to me after almost sixty consecutive years of life in the United States. I still find it difficult to call people by their Christian names unless I have known them well for a long time or they are much younger than I, and I am still embarrassed at being called René by young people with whom I have shared only limited activities. Radio and television programs may now serve the role played in the past by fables. And it has been my experience that while these programs have much in common all over the world, there are differences of emphasis and interpretation among them from culture to culture.

Regardless of the characteristics of the natural environment, one can thus find everywhere in the world an immense diversity of cultures and of social systems based on as many different images of humankind created through the processes of socialization, especially during childhood. The assumptions from which these images are derived usually remain quite stable for a long time, but they are not unchangeable.

Since the sixteenth century, for example, the image of humankind has been affected by the Copernican revolution which showed that the earth is not in the center of the cosmos as used to be believed. More recently the Darwinian revolution has revealed that *Homo sapiens* is just one expression of a universal evolutionary process which began billions of years ago and is continuing now. In my judgment, one has exaggerated the effect of these scientific revelations on the mind of the general public. Many sophisticated scholars living in Greece or China 2500 years ago, or in western Europe during the Renaissance or the Enlightenment, probably had images of themselves and of their relation to the cosmos not very different from that of an American or European scholar today. As to the persons in the street, I doubt that they are much more concerned with such problems now than was an average citizen anywhere in the Western world centuries ago. Under usual circumstances most of us still behave and feel as if we were at the center of the world.

There is no doubt, on the other hand, that the image of humankind has been profoundly influenced by scientific technology. Few are the people in the countries of Western civilization who do not take it for granted that we can, if we try hard enough, achieve mastery over most of the forces of nature and use machines to replace almost all types of human work—mental as well as physical. We are prone to regard ourselves as superior to the rest of creation and qualitatively different from other animals. Furthermore it has been suggested that we shall eventually improve on *Homo sapiens* by marrying the human body and mind with machines. This could be done, it is said, by implanting computers and other sophisticated devices into the human organism so as to improve and modify at will its physical and intellectual capabilities. The hypothetical man-machine complex which would thus be created has been called "cyborg" (the abbreviation for cybernetic organism) and would permit a two-way communication between the biological and the mechanical components of the system. An extensive development and use of cyborgs might be regarded either as the mechanization of humankind or the humanization of the machine. In any case, its inevitable consequence would be a profound change in our image of humankind but I am not much concerned about this prospect because I doubt that many human beings would be happy functioning in a cyborg civilization.

Until a few decades ago, most people believed that the more extensively and intimately we introduced sophisticated machines into our lives, the more this would contribute to progress. There is at present a feeling, however, that the changes thus brought about are not all to the good. Western technology is accused of creating conditions which are inimical to health, to imagination and to the quality of social life and of the environment. The disenchantment with technological civilization may not be sufficiently deep to bring about a significant retreat from present life-styles but it is widespread enough to make most people of Western civilization look with favor on ways of life which are less machine oriented than they are now.

Granted that the images of technological man and the cyborg way of life have lost much of their glamor and that our societies will try to recapture some of the values of Arcadia, technological imperatives will nevertheless continue to prevail. They will probably lead to super-systems designed to help people in the development of products, analyses and decisions. These supersystems may eventually, if they have not already done so, reach such a state of complexity that they cannot be fully understood by their users who will nevertheless find them indispensable in order to continue operating in our modern societies even at the risk of unpredictable dangers. Thus, while the environments we are in the process of developing are important for our own daily lives their greater importance is that they constitute the program from which will emerge the printout that will determine the formula of life we transmit to succeeding generations. Our ever-increasing involvement with machines contributes to the image of humankind we create and thereby profoundly influences the future orientation of society.

CHOICES AND CREATIVITY

The view that man can shape the future through conscious decisions concerning his environment was picturesquely expressed by Winston Churchill in 1943 while discussing the architecture best suited for the Chambers of the House of Commons. The old building, which was uncomfortable and impractical, had been almost bombed out of existence during the Second World War. This provided an opportunity for replacing it with a more efficient one, having greater comfort and equipped with better means of communication. Yet Mr. Churchill urged that the Chambers should be rebuilt exactly as they used to be. In a spirited speech, he argued that the style of parliamentary debates in England has been conditioned by the physical characteristics of the old House, and that changing its architecture would inevitably affect the manner of debates and, as a result, the structure of English democracy. Mr. Churchill summarized the concept of interplay between man and the total environment in a dramatic sentence that has general validity for human life: "We shape our buildings, and afterwards our buildings shape us."

In his speech Mr. Churchill clearly regarded the House of Commons less as a physical building than as an expression of British society. His statement that our buildings shape us referred to the maintenance of certain parliamentary practices, but it applies just as well to all the characteristics that define a nation, a social class or a way of life. The fact that we are profoundly influenced by the environments we create is rather frightening since it appears to imply that we are the helpless pawns of deterministic forces. Fortunately, as Mr. Churchill's own life demonstrates, and as I shall repeatedly emphasize in this and the following chapters, we have a great deal of freedom in choosing and modifying our environments. We can even, if we desire, transcend their effects.

In their scientific studies, biologists tend to ignore that human development can be markedly influenced by our ability to govern our individual lives. Geneticists deal with the mechanisms through which the genes we inherit from our parents direct everything we do and become. Environmentalists emphasize that we are shaped by our surroundings and by the events we experience. Behaviorists would have us believe that we are hopelessly conditioned, even beyond freedom and dignity. All attitudes about the deterministic aspects of human life can be supported with a large body of facts, but scientists commonly overestimate the explanatory value of scientific knowledge. They seem to suffer from a kind of parochialism common among specialists, namely the belief that human life can best be explained by the phenomena studied in their own professional specialty. Human nature, however, is not so simple that it can be reduced to the knowledge available to twentieth-century scientists. As mentioned earlier, for example, the existence of free will cannot be demonstrated, let alone explained scientifically, but this failure does not weigh much against the commonsense observation that human beings, and most likely animals also, constantly make choices and take decisions.

We do not usually remain passive when we realize that certain situations may have unfavorable effects on our lives. Most of us have some degree of freedom in moving away from places that we consider unsuitable and in seeking others that we consider more desirable. Moreover, we can actively respond to our surroundings in a manner which is often original and creative, and thus impose the direction of our choice on our own development. The freedom to move and to change is basic to the processes of self-discovery and self-realization. In fact, the ability to imagine possible futures and to create our persona out of the options opened to us is a human attribute that becomes manifest very early in life.

At the time of birth, infants might be regarded simply as little animals, but they soon transcend their biological endowment by acquiring the cultural heritage of the group in which they grow. Infants are aware of their environment and store information about it from the very first day of life. They soon exhibit individuality in their patterns of response to what they experience; far from being passively conditioned by stimuli, they behave very early as searching participants in the learning process. This initial phase of psychological maturation is followed by more and more conscious activities through which children create their personas from the uniqueness of their genetic constitution and early experiences. In midchildhood, perhaps by the age of five, most children have acquired enough environmental information and have sufficiently developed individual patterns of responses to imagine a world of their own in which they can act out their personas.

Human development continues to the extent that the person learns to deal creatively with a greater and greater diversity of stimuli and to take initiatives according to a system of values, which is personal, even if it is but a modification of the value system of the social structure. These values include anticipations of the future that have their roots deep in the past but are also the expression of our own tastes. It is because we are genetically endowed with the ability to imagine, to symbolize, to anticipate the future and to choose among options that we can create the physical and conceptual environments in which we spend our lives.

As outlined above, the process of self-creation implies the existence of a wide spectrum of conditions among which persons can operate to develop their own personas and to act out their own way of life. However, freedom is an empty word if the opportunity to choose among different options does not exist. For example, children born and raised in urban slums are theoretically free, but their range of choices and opportunities to move are commonly so limited that they may find it difficult (although never impossible) to overcome the forces of environmental determinism. Children growing in economically prosperous classes also may suffer from environmental deprivation if their surroundings are deficient in human values. Emotional and experiential poverty is not uncommon in wealthy and polished surroundings, such as those of many stereotyped suburban settlements.

From the point of view of human development, the diversity of the environment is thus of greater importance than its comfort, its efficiency or even its esthetic quality. The loss of diversity that occurs when houses with individual tastes are replaced by buildings of monotonous character usually results in such sensuous impoverishment that William H. Whyte was once moved to plead for "at least one hideous house to relieve the good taste."

By providing a wide range of options, environmental diversity helps us to discover what we like, what we can do and what we want to become. But self-discovery is frustrating without self-realization, and this requires physical, mental and emotional involvement in some cause. Human development implies therefore that, in addition to environmental diversity, the person can participate actively in events instead of merely watching them as a passive spectator. Skills in intellectual activities or in human relationships do not develop any better while viewing a television program than do muscles develop while watching a ball game.

Human development naturally proceeds through the orderly unfolding of processes encoded in the genetic constitution, under the influence of environmental forces, but this does not mean that the responses of the organism to stimuli are blind and passive expressions of biological mechanisms. In most cases, indeed, development is influenced by deliberate choices and by anticipations of the future.

Admittedly, a large percentage of responses are determined by instincts that operate outside consciousness and free will. Instincts enable us to deal in a decisive and often successful manner with life situations similar to those repeatedly experienced by the human species in the evolutionary past, but instincts are so precisely pointed and mechanical that they are of little use for adaptation to new circumstances. Yet this kind of adaptation is essential to continued development.

Whereas instincts provide for biological security in a static world, awareness, knowledge, and motivation account for the creativity of human life. To the extent that we can make choices, we can give a direction to our adaptive responses and thus influence our development. We can make choices concerning our life-styles and our surroundings so as to favor the development of bodily and mental attributes that we judge desirable. Our thought processes also can affect our development by influencing our hormonal and other physiological mechanisms—let alone the intellectual and mental attitudes that are reflected in all aspects of our lives.

Just as choices and decisions affect normal development, so they also influence reeducation designed to correct disabilities either of innate nature or resulting from accidents or pathological processes. Development implies more than the passive unfolding of genetic potentialities and so does reeducation involve more than passive training. In both situations, the organism must participate as a whole in activities selected to foster a truly creative process of adaptation and growth.

It is obviously easier to follow one's instincts passively than to govern consciously one's creative responses; hence the worried expression on human faces at the time of decision. In the words of the theologian Paul Tillich, "Man becomes really human only at the time of decision." Being human implies the willingness to make the efforts required for growth through creative adaptations.

For animals, as mentioned earlier, a good life means carrying out the kinds of activities for which they have been conditioned by their genetic constitution and their early experiences in their natural habitats, but this will not do for human beings because most of us now live in environments to which we are not biologically adapted. Throughout historical times and even prehistory, furthermore, individual persons and whole cultures have taken risks that had no apparent biological justification—to conquer land or acquire wealth, to explore unknown regions or to solve a scientific problem. There seems to be little in common between Alexander the Great's urge to master the world, the determination of the physicians Carroll and Lazear to understand the cause of yellow fever by experimenting on their own bodies, Mallory's desire to reach the top of Mount Everest, and my naive eagerness to experience America because of childhood fantasies about Buffalo Bill and the Far West. Yet, there is a common component in all these seemingly unrelated goals, namely a desire for physical or mental adventure. This desire, which seems to be one of the invariants of human nature, can take many different forms and can result in different types of creativity.

There are many situations, of course, in which human behavior is dictated by biological necessities; people need food, shelter, space and comfort just as animals do, but most human activities, as already mentioned, do not seem to have an obvious biological utility. From this point of view, the behavior of the human species became different from that of animals as far back as the Old Stone Age. All over the earth and in all prehistoric and historic periods, human groups have devoted an enormous percentage of their resources, energy and imagination to enterprises that had little relevance to the biological needs of *Homo sapiens,* considered as an animal species.

The population of France and Spain did not exceed 50,000 people during the last Ice Age, yet it created an immense number of artifacts of great complexity and artistic quality which must have had great symbolic significance in the life of these people—even though we do not fully understand this significance. The statuettes known as paleolithic Venuses, the complex and spectacular cave paintings, the countless objects of stone, bone and ivory with detailed and exquisite designs, all played some important role in Cro-Magnon life, but certainly were not essential for survival. Similarly, minuscule populations were involved in the erection of the immense megalithic structures such as the Stonehenge circle in England, the Carnac alignments in French Brittany, the gigantic statues of Easter Island. And so it goes throughout prehistory and history. Everywhere and always a very large percentage of human energy and imagination has been and still is devoted to activities that seem irrelevant to the purely biological needs of *Homo sapiens.*

Many of the great creations of early historical times were, admittedly, achieved through slave labor, as was the case for the Egyptian pyramids, but others were community enterprises which had an intense social or spiritual meaning for the people involved. Most of the European monasteries and cathedrals of the Middle Ages were erected in towns of less than 10,000 inhabitants—even Paris had only 35,000 inhabitants when Notre Dame was started—and the same is true for the Renaissance palaces and religious buildings of Italy. In our times, also, the space program of the 1960s required a large percentage of the national budget in the United States and the U.S.S.R., a sacrifice which was willingly accepted by the majority of the American and Soviet populations.

We create artifacts of no compelling biological utility because, as mentioned earlier, we are *in* nature as other animals are, but no longer quite *of* nature. We are rarely if ever satisfied with watching nature passively; and indeed could not do it if we tried. Even when we do not touch nature, we develop feelings about it and engage in mental reconstructions of it in which we incorporate much of our personality. We cannot just see a landscape; we add to it a certain mood, the outcome of which is a picture that we either simply imagine or convert into a painting on a canvas. We cannot just listen to natural sounds; we incorporate them into music, whether in our mind or on a score. We demand of a dwelling that it be more than a shelter; however primitive, we make it a home fitting some personal or social wants that transcend biological necessity. For human beings, survival is not enough and we cannot avoid intervening mentally and physically into nature. Creativity has often introduced into human life complexities for which we are not biologically prepared and which have generated in the past as well as now problems that may lead to disasters, but concern about consequences is usually less powerful than the urge to choose and to create and can only serve as a guide for safer and better creations.

SELF-DISCOVERY

My mother and my schoolteacher were the two persons who influenced me most during the twelve years of my village life in the Île de France. From then on, the most important aspects of my social conditioning came from my contacts with the general public in the streets and parks of Italy, England and the United States—but first and especially in Paris.

In the village where I grew up I remember distinctly a conversation with my mother one evening while I was helping her to wash the dishes in the kitchen when I was approximately ten years of age. As she frequently did, she expressed her desire that I move into a broader life than what she had known, not only more prosperous but also more intellectually rewarding. My mother had received little formal education, having left school to work as a seamstress at the age of twelve, but this was a time when primary education was excellent all over France. As she was extremely sensitive and perceptive, she had learned a great deal, and was eager to prepare me for a brilliant future. Late in that particular evening, she opened the *Petit Dictionnaire Larousse,* the only learned book in our home, at the special section devoted to the "Grandes Écoles." This provided both of us with material for daydreaming as to what my future should be. I do not know whether she had a clear notion of what the Grandes Écoles stood for but it is certain that my attempts at a life of scholarship have their origin in her attitude, not only on that particular evening, but throughout my teenage years, even while we were in great financial difficulties after the death of my father at the end of the First World War. Whenever I have been successful as a scholar or otherwise, there comes to my mind the pink pages at the end of the Larousse dictionary where, for the first time, I read the brief descriptions of the Grandes Écoles and thus obtained a somewhat concrete image of a world larger in scope and more sophisticated than the one in which I had lived.

I also remember with much gratitude Monsieur Delaruelle, the dedicated and thoughtful teacher who ran alone the one-room school for the boys of the village. We were approximately fifty boys from age five to twelve and Monsieur Delaruelle taught us, with the same enthusiasm, all aspects of knowledge from arithmetic, to grammar, to history and music. In addition to conventional teaching he did his best to acquaint us with events of importance then occurring in the world and to make us aware of their significance. On a certain day—it must have been around 1911—he reported to us with great excitement that Leonardo da Vinci's *Mona Lisa* had been stolen from the Louvre, and took occasion of this piece of news to tell us about art.

Living in a village or small town had certain educational advantages at the turn of the century. Children had the opportunity to observe various trades at close range and not only to participate in them, but to take responsibilities. Such direct observation and participation gave a more concrete knowledge of reality than can be obtained in large cities even equipped with the most up-to-date museums. This was especially true in the past because small towns and villages were then essentially self-sufficient.

Being raised in fairly small human settlements may furthermore contribute to the self-confidence of children by giving them a greater sense of their own importance relative to the rest of the world than they are likely to have in an urban agglomeration. Until the age of thirteen, I had lived only in two villages, both of less than five hundred inhabitants and in a town, Beaumont-sur-Oise, the population of which probably did not exceed 3,000. Thus, I never was less than $1/500$ or $1/3000$ of the human world in which I functioned. Furthermore, I could know the human beings among whom I lived in a way different from that children now experience in large cities. I saw people of all ages not only at home but also when they exhibited the human foibles that manifest themselves during any normal life.

The human atmosphere of the village or the small town still had, in my youth, the demographic, psychological and emotional dimensions of the tribal structures in which humankind lived until the agricultural revolution 10,000 years ago, and of the villages in which the immense majority of the human population has lived, on all continents, until our times. The most important problem of urban planning may well be to recreate, within our large cities, the equivalent of the diversified unit of a few hundred people in which the social evolution of humankind took place, and to which we are still adapted today.

My environmental conditioning naturally changed character when I moved to Paris, where it became somewhat less deep and less personal but more varied. Parisians walked much faster than people in my village and with a different rhythm. Much later I experienced the difference in rhythm of walk not only between the French countryside and Paris but between Paris, London, Rome and New York. Each great city has its own rhythm of walk which certainly corresponds to differences in general attitudes toward life. Even more important than the physical stimulation caused by change of rate was the mental stimulation which came from the fact that, in Paris, streets differed from each other; each "arrondissement," furthermore, had many different facades and human moods with as many invitations to adventure. I had become aware of the immense diversity of people from fables, from history and stories, and from my limited personal experiences. It was only when faced with extreme diversity in the different sections of Paris, however, that I began to know envy. I recognized then that many aspects of public life were unavailable to me, simply for economic reasons. My first painful awareness of this limitation was when I walked by a theater on the opening night of a new play. The fables I had read had expressed the vanity of such occasions, but their message was forgotten when I realized that I had very little chance of ever being financially able to participate in this fashionable life.

The public squares and parks of Paris were the places where I completed my conditioning as a youth. My father's shop was located on a street next to small factories. As was the custom in those days, men and women would sing the new popular songs at the corner of the streets or in the public squares and sell copies of them for a few pennies while the public repeated the words and tunes. I was so much a part of the scene that I can still hum most of these songs from memory. I also attended the public fairs at the beginning of my Paris life but soon lost taste for this kind of entertainment. In contrast, I became more and more fond of the various parks of Paris, each with its own distinctive personality.

There were the small parks and walking areas along the Seine where children with their parents and grandparents played during the day hours, and where teenagers as well as adults of almost any age engaged in the games of love at almost any hour of the day or night. There was the austere Jardin des Plantes, with its ancient trees, exotic plants, statues of famous naturalists and the awareness that its buildings sheltered a rich history and promoted mysterious scientific knowledge. There were in particular the Parc Monceau not far from the Lycée Chaptal where I was a student from age thirteen to eighteen and the Jardins du Luxembourg (more commonly called Parc du Luxembourg) close to the Institut National Agronomique where I spent two years between 1918 and 1920. These two parks were of special interest for me because I spent many hours on their benches reading for study or for pleasure, and perhaps more usefully daydreaming while observing various kinds of people.

The Parc Monceau and the Parc du Luxembourg were then and still are profoundly different in physical appearance and human atmosphere. The Parc Monceau was designed in the late nineteenth century under the influence of the English so-called "natural" landscape style; it is located in one of Paris's most fashionable districts and the public in it consisted then chiefly of persons of wealth, or at least assuming the airs of the wealthy. Watching them suggested to me an opulent way of life very different from mine and that I imagined highly sophisticated and enjoyable. The Parc du Luxembourg in contrast has a classical design which reveals its seventeenth-century origin. Being located in the Latin Quarter near the Sorbonne and several of the Grandes Écoles, it was always full of young people reading or more likely arguing in an intense and serious mood as they walked to and fro along the straight allées. The Fontaine Médicis, a small artificial pool with a waterfall and shaded by ancient trees, seemed completely removed from the brighter and busier sections of the Parc and was always surrounded by romantic souls in search of poetic silence for their meditations.

As time went on, the Parc Monceau and the ways of life it symbolized lost much of their appeal for me. The Parc du Luxembourg, in contrast, provided me with a spiritual climate that I found increasingly exciting, even though I still had only very faint ideas concerning the world of learning with which it was associated. This choice between two contrasting atmospheres was the first and probably the most important one of my life—but it had not been entirely mine. Its origins were in my mother's hopes of scholarship for me and in Monsieur Delaruelle's dedication to teaching.

While the early influence of my mother and of my schoolteacher had prepared me emotionally for the belief that a satisfactory future would depend on my willingness to study, it was my direct perception of life in the parks and the streets of Paris that enabled me to become aware of my personal tastes. I soon lost my desire for certain opulent life-styles that had first seemed highly desirable to me; I remained jealous of the people who could afford them, but not envious of what they did with their money. On the other hand, I came to realize that I was attracted, even though in a vague way, by activities and social atmospheres that were more subdued than glamorous social life, but that appeared nevertheless more rewarding. I had been shaped at first by the peaceful environment of the Île de France villages; I was born to a new life of intellectual adventure in the Parc du Luxembourg.

My own experience fits with the teaching of history that one of the most valuable aspects of urban life is to help people discover what they would best like to do and to become. Providing the right conditions for this process of self-discovery has much relevance to the design of urban agglomerations. New York City is at least as rich as Paris, London, Rome or any other world metropolis in diverse and stimulating environments, but its design seems to me less favorable for self-discovery than is that of great European cities. Children raised in Harlem or Bedford-Stuyvesant find it much more difficult than it was for me to experience the various sociocultural atmospheres of the city and thus to discover by themselves the life-styles they prefer; and the same difficulty applies to many children of more prosperous sections of Queens, the Bronx, and Staten Island.

There comes to my mind an occasion a few years ago when Rockefeller University acted as host to several members of the National Academy of Sciences who served on a committee appointed to investigate the conditions in Harlem. As they returned from their survey, one of the academicians told me that he had been prepared for everything he had seen—the poverty of people, the dilapidated buildings, the dirt in the streets—but not for seeing children's faces against the windows, with nothing to look at but squalor and rubble. The children of Harlem and Bedford-Stuyvesant have full freedom to go where they want and to act as they will, but freedom of choice is an empty word if a wide variety of options are not available from which to choose.

Since the thirst for adventure, rather than a basic antisocial attitude, is probably often responsible for antisocial behavior, increasing environmental diversity might help to decrease social misbehavior. Selecting the kind of adventure suited to one's personality is an essential factor of development through self-discovery. In this light, environmental diversity, with various options available to people and especially to children, is an essential aspect of true functionalism. Most American cities are dysfunctional in this respect.

We are unquestionably shaped to a very large extent not only by our genetic constitution but also by the environments in which we function and by our ways of life. For this reason it is of extreme importance that we have as much freedom and as many options as practically possible in selecting or creating the environmental conditions that shape us. In the final analysis, human development is determined less by the forces to which we are passively exposed than by the choices we make concerning our personal lives and the organization of our societies.

3

THINK GLOBALLY, BUT ACT LOCALLY

LOCAL SOLUTIONS TO GLOBAL PROBLEMS

THE GLOBAL VILLAGE

THE NETHERLANDS, HORIZONTAL COUNTRY CREATED BY HUMANKIND

MANHATTAN, THE VERTICAL CITY

THE GLOBE-TROTTER AT HOME

3

THINK GLOBALLY, BUT ACT LOCALLY

LOCAL SOLUTIONS TO GLOBAL PROBLEMS

On American and Canadian college campuses where I recently lectured, most students were intensely concerned with environmental and social problems, but chiefly in their large-scale aspects, and preferably at the national and global levels. Faculty as well as students were surprised and somewhat annoyed when I suggested that, instead of being exclusively concerned with the nation or the world as a whole, they should first consider more local situations, for example, the messiness of public rooms on their campus and the disorder of their social relationships. My message to them was that thinking at a global level is a useful and exciting intellectual activity, but no substitute for the work needed to solve practical problems at home. If we really want to contribute to the welfare of humankind and of our planet, the best place to start is in our own community, and its fields, rivers, marshes, coastlines, roads and streets, as well as with its social problems.

I had many occasions to ponder on the local aspects of global problems while participating directly or indirectly in the huge international conferences organized during the 1970s under the auspices of the United Nations to discuss the contemporary problems of humankind. There was a pattern common to these megaconferences. They all began with resounding statements of global concern and with clarion calls for international thinking and action. As the meetings went on, however, discussions of concrete issues soon became hopelessly diluted in a flood of ideological verbiage unrelated to practical action. At the end of the conference, the efforts to set down a statement of consensus yielded resolutions so broad and so vague in meaning that few of them could be converted into action programs. As a result of these observations, I came to believe that such international conferences are a waste of time, but I have now changed my mind, for two different reasons.

On the one hand, the megaconferences of the 1970s generated a global awareness of certain dangers that are now threatening all nations, the rich as well as the poor. This is not a small achievement because thinking globally is not easy for human beings. As a species, *Homo sapiens* has evolved in small social groups and in limited physical environments so that our intellectual and emotional processes are not biologically adapted to global or long-range views of any situation. It is only when people from all parts of the world have the opportunity of listening to each other's problems that they realize, even though with difficulty and slowly, how crowded we are on our small planet, how limited are its resources, and how multifarious are the dangers to which we are all increasingly exposed.

The megaconferences of the 1970s had the additional merit of bringing to light the diversity of physical and social conditions on our planet and to dramatize the consequences of this diversity. While there was much posturing during the conferences, the official delegates learned from representatives of other countries that global problems appear in a different light depending upon local situations. The environmental purists of the Western world discovered, for example, that abject poverty is the worst form of pollution and that many poor countries have legitimate reasons to be more interested in economic development than in the ecological gospel. At the 1976 U.N. Conference on the Habitat in Vancouver, the poor nations naturally complained of being exploited by the rich industrialized countries but they also realized that they had much to learn from Western civilization concerning advanced technologies applicable to such problems as water supply, low-cost housing or sustainable rural development—let alone industrial development.

The most valuable achievement of the international conferences was probably, however, to reveal that the best and commonly the only possible way to deal with global problems is not through a global approach but through the search for techniques best suited to the natural, social and economic conditions peculiar to each locality. Our planet is so diverse, from all points of view, that its problems can be tackled effectively only by dealing with them at the regional level, in their unique physical, climatic and cultural contexts. Three examples will suffice to illustrate the necessity of the local approach to global problems.

• The recommendations of the Vancouver Habitat Conference were explicit with regard to the fact that all people need clean water and decent shelters. The techniques required to meet these obvious biological necessities, however, must be designed to fit local conditions such as the density of the human settlements, the topographical, geographical and climatic conditions, and of course the economic resources. The design of shelters is further complicated by local social habits and tastes. The recommendations concerning cultural matters or quality of life had to be even less specific because these values have intense local and historical characteristics that transcend scientific determinism and definitions.

• The word "desertification" refers, not to natural deserts, but to areas which are being rendered desertic by human activities, especially by overgrazing and by the use of wood as fuel. Since desertification is a problem of increasing gravity in many parts of the world the United Nations Environment Programme (UNEP) first attempted to control its spread with projects which were transnational, in the sense that they dealt with vast continuous areas stretching across several countries. However, this transnational approach had to be abandoned because the social and agricultural practices leading to desertification differ from country to country. The desert unit of UNEP has recently decided that, before receiving international help, the individual countries must formulate their own projects fitted to their particular agricultural and social practices.

• Until 1973, the low cost of petroleum and natural gas, and the ease with which these fuels could be shipped and used anywhere in the world, created the illusion that fairly uniform technological policies could be formulated for the planet as a whole. However, petroleum and gas are becoming much more costly and soon will be in short supply. For this reason, plans are being made to replace them by coal, which is much more plentiful, and eventually by different kinds of renewable sources of energy such as nuclear fission (and perhaps fusion), solar radiation, the wind, the tides, the waves, and the different kinds of organic materials grouped under the name of biomass. Each one of these sources of energy has advantages and objections peculiar to it and, unlike petroleum and gas, each one is much better suited to one natural or social situation than to others. For example, coal is not available everywhere; its shipment over long distances is costly and its mining results in types of environmental degradation that differ from region to region. Solar radiation has a better chance to be developed on a large scale in regions of intense insolation, the wind where it blows in a fairly dependable manner, the biomass in densely wooded areas, nuclear fission in industrialized countries that are deficient in other energy resources and where the public is therefore more likely to accept the risk of massive unpredictable accidents.

Just as the shift from hydroelectric power to coal, then to petroleum and gas, made certain heavy industries move from New England to the Appalachians and then to Texas, so we can anticipate that future industrial developments will differ from one place to the other when local solutions to the global energy problem have been worked out.

In my opinion, it is fortunate that practical necessities will compel different local solutions to global problems. Globalization inevitably implies more standardization and therefore a decrease in diversity, which in turn would slow down the rate of social innovations. Another danger of globalization is that excessive interdependence of systems increases the likelihood of collective disasters if any one of the subsystems fails to function properly as a result of accident or sabotage. Finally, we may soon reach a point, if we have not reached it already, at which the technological, economic and social systems become so huge and so complex that they cannot be readily adapted to new conditions and cannot continue to be really creative. The human mind cannot cope with the comprehension, let alone the management of systems which are too large or too complex, even when these are of human origin. In contrast, there is a better chance for adaptability, creativity, safety, and manageability in multiple fairly small systems, aware and tolerant of each other, but jealous of their autonomy.

Skepticism concerning the value of globalization does not imply isolationism. The ideal for our planet would seem to be not a World Government but a World Order, in which social units maintain their identity while interplaying with each other through a rich communications network. This is beginning to happen through the specialized agencies of the United Nations (there are at least sixteen of them at the present time) such as the World Health Organization, the World Labor Organization, the Food and Agricultural Organization, the World Meteorological Organization, and the UNEP which I mentioned earlier. Their existence and success justify the hope that we can create a new kind of global unity out of the ever-increasing diversity of social structures.

Focusing on a local problem is thus very different from retreating into isolationism. In fact, it necessarily requires the operation of several kinds of social networks involving scientists, industrialists, politicians and private citizens. Controlling water pollution in the Great Lakes is making some progress through the enactment of multiple complex agreements at industrial and political levels between the United States and Canada. Similar progress has also been made in the case of the Rhine to the point that the four nations involved—Switzerland, France, Germany and Holland—have formulated and actually put into action a system of fines to be paid by the industries that pollute the river. The fate of the Mediterranean appeared hopeless a few years ago but now, after decades of incredibly complex negotiations involving all the Mediterranean countries, there is some hope that this highly local problem can be progressively solved through agreements concerning the control of domestic and industrial discharges.

THE GLOBAL VILLAGE

Most of human life since the Old Stone Age has been spent in small, fairly stable communities, consisting of either moving bands of nomads or small stable villages, which were organized to satisfy the invariants of humankind out of the resources locally available. There have been countless political revolutions and other upheavals in the course of history but their final outcome has always been to recreate communities of a few hundred to a few thousand people in which everyone knew his or her place in the social order of things and accepted, willingly or under duress, the local rules of the game. Surprising as it may seem, this pattern of social structure still prevails to a large extent in most of the world today. The word "community" or its equivalents in other languages has everywhere a deep sentimental appeal. And the commune is the fundamental social unit in the People's Republic of China.

In *The Colossus of Maroussi,* Henry Miller vividly conveys his belief that, at the time when Knossus was the political center of ancient Crete, the island was a gay, healthful, sanitary, and peaceful community. According to him, this happy attitude was maintained under almost any form of government—at times under the domination of Imperial Egypt, later under the homely immediacy of the Etruscan rule, and still later through a communal spirit of organization similar to that of the Incas. Miller felt that there was "something down to earth about Knossus"—

the influence of religion was graciously limited and women played an important role in public affairs; human beings were "religious in the only way that is becoming to [them] by . . . extracting the utmost of life from every passing minute. Knossus was worldly in the best sense of the word." In contrast, still according to Miller, the fundamental weakness of our own civilization is the absence of anything approaching a communal existence. Yet it seems from many of Miller's writings about other European areas and from my own personal experiences that various forms of communal existence still provide today the basis for much of human life in most parts of the world.

Many crowded areas give the impression, admittedly, that the community structure is breaking down or is of ever-decreasing importance, in poor as well as in rich industrialized countries. For example, the problems of India immediately call to mind not communities but Calcutta, Bombay, Delhi, with their huge populations and tragic slums. This image also fits most developing countries each with its own kind of shantytowns, favellas, barrieros and bidonvilles. The fact is, however, that most of these slums have a social structure consisting of fairly well-defined small communities. Furthermore, emphasis on huge cities gives a very inaccurate picture of human life on earth. The immense majority of human beings in poor countries do not live in large cities but in millions of small villages, each of a few hundred people on the average. This is true even of India where more than 600,000 such villages account for the very largest percentage of the total population.

At first sight, the demographic and social picture in the poor countries appears to bear no relation to the state of affairs in the countries of Western civilization, most of which are heavily urbanized and are in the process of becoming more so. Many parts of the industrial world have gone beyond the phase of metropolis to that of megalopolis such as illustrated in the United States by the uninterrupted urban area along much of the Atlantic seaboard, the Pacific Coast, and the Great Lakes. Similar conurbations exist in Europe and Japan. The Greek planner, C. A. Doxiadis, believed indeed that megalopolis would progressively evolve into Ecumenopolis—a single continuous world city.

Ever since the late nineteenth century, there have been a variety of plans to develop artificial urban systems having little in common with the traditional city. The concept of an entirely new type of urban development seems to have been formulated first by the Spanish engineer, Soria y Inata, who proposed around 1880 that a continuous linear city be created by extending existing settlements along major routes of transportation. This idea, which was revived in 1910 by the American engineer Edgar Chambless in his book, *Roadtown,* has been given a grim practical expression in some of the Russian settlements of the 1930s. *Roadtown* has spontaneously developed along many American highways, destroying almost everywhere the coherence of the towns and villages that stand in its path. Many other projects have been formulated to develop "instant cities," "plug-in cities" and other such purely technological urban agglomerations—which all have in common the ultimate goal of replacing traditional cities by urban "throw-away" settlements, which in their turn would be replaced at suitable intervals by others better suited to the technologic and economic possibilities or needs of the time.

Many people, on the other hand, have come to feel that linear cities, plug-in cities and other forms of instant cities are not for the sane; they believe that a good human life cannot be experienced in urban settlements designed as if they were indeed disposable containers for dispensable people. During the past few decades, there has been increasing acceptance of the idea that the human settlements of Western civilization must be restructured on a human scale to facilitate and enrich the human encounter. Whether fully successful or not, conscious efforts in this direction were first made in the British New Towns and since then in the many new towns which have been or are being created in several parts of Europe and to a smaller extent in the United States.

There are at least twenty-four new towns in Great Britain and eight in France. They range from 30,000 to 200,000 inhabitants but in all cases they are planned in such a manner as to be subdivided, by various artifices, into much smaller subunits that correspond to neighborhoods having a marked degree of social self-sufficiency, with their own schools, shops, taverns, recreation grounds and a certain measure of administrative autonomy. Even in Amsterdam, where the planners are committed to increase the size of the city up to one million inhabitants, the plans of development are essentially similar to what they were a generation ago, namely, neighborhood by neighborhood, each equipped to function as a fairly complete social unit. Venice was planned from the very beginning as a city of neighborhoods made up of parishes related to a particular church and a square, and it has retained much of this human scale.

It is said that London, despite its enormous size (it was the first city to exceed one million population in the nineteenth century), nevertheless remains in many respects an assembly of villages. As to Paris,

I know from personal experience and from many decades of contact with family and friends, that the highly centralized administration has not decreased the importance of neighborhoods, each with its social characteristics and often with identifiable architectural features. The Parisian neighborhood is not just a postal or political district but the expression of strong historic and social forces. The sense of belonging to one particular *quartier* or even *arrondissement* is just as intense in the apartment dweller, the shopkeeper or the bistro customer as is the sense of being a Parisian. I can also write from personal experience about New York City. Even though the gridiron pattern of Manhattan would seem to be an obstacle to the emergence of neighborhoods, Yorkville, Chelsea, Greenwich Village, SoHo, the different sections of the West Side nevertheless exist as distinctive entities, not because they have characteristic architectures, but because they are identified with certain ways of life and therefore attract certain kinds of people. And the same applies to the other four boroughs of New York City. In his book *New York* the celebrated French writer and globe-trotter Paul Morand referred in 1930 to "le Bronx amorphe et anonyme." Little did he realize the extent of the physical and social differences between Riverdale, the Grand Concourse, the South Bronx and Pelham Bay, all within the Bronx borough.

Thus, the neighborhood is a social fact even in the largest and seemingly most anonymous settlements, in the rich as well as in the poor countries. The neighborhood exists in an inchoate form even when it is not articulated in an official plan or provided with the institutions that would make it an identifiable administrative community. Neighborhoods emerge not only because people sharing the same place cannot avoid developing common interests, but even more because the very characteristics of the place, whether physical or social, attract people who have at least some tastes and interests in common and therefore tend to join in common public enterprises.

The general increase in mobility and the widespread use of technology by more and more standardized methods certainly contribute to the homogenization of human life and give the impression that the advocates of One World speak for the future. While studying Indian reservations and Eskimo communities in the mid-1970s, the American psychologist Robert Coles had the occasion to observe that the penetration of the "American ways of life" was not limited to such artifacts as Coca-Cola, pizzas, hi-fi sets and snowmobiles but had reached into the psychologic manipulation of people. Dr. Benjamin Spock's book,

Baby and Child Care, was common in certain Eskimo settlements. Dr. Coles found the following message printed on mimeographed handouts on the Hopi reservation: "Something on your mind? Don't be silent. . . . Come, talk and feel a lot better afterwards." An Indian bulletin board near the Rio Grande publicized a pamphlet with questions and advice: "Unable to sleep? Overweight? Having marriage problems? Come talk about it! You'll feel better." These exhortations are the more surprising because Eskimo and Indian peoples are not prone to express their feelings to strangers. Although Eskimo and Indian settlements are thus exposed to patterns of social behavior common in American life, there is as yet no convincing evidence that publicizing these patterns has significantly affected their everyday life.

One can take it for granted, in any case, that whether homogenization is technological or psychological, it will chiefly affect limited aspects of existence and of the environment—for example, means of transportation, facilities for travelers, safety of air, of water and of food, and services common to large groups of people. Whatever the ethnic group, the traveler will increasingly move through repetitive networks of thruways and airlines to reach similar mazes of lobbies, elevators, bedrooms, snack bars, restaurants—indeed of healers and palm readers—at the point of destination. Such uniformity will be boring but will help travelers to orient themselves in a new place and to function in it comfortably or at least easily, thus sparing energy for discovering and enjoying what is new or interesting to them wherever they have to stop.

A powerful trend toward uniformity is thus obviously present throughout the world with regard to those aspects of life common to all human beings, but there is simultaneously an opposite trend toward regionalism. The One World of the future will be made up of many different local worlds because the quality of our lives depends largely on emotional, esthetic and spiritual satisfactions that result from the contacts each one of us makes with our physical and social surroundings. The very trend to uniformity and the sense of boredom it connotes tend to make many of us interested in the unique characteristics of the place where we live, in its traditional styles and in the still unexplored possibilities it offers. In my judgment, the cultivation of the sense of place will increase in importance as more of our very public activities and experiences become globalized.

The desire to be different may well play a role in the worldwide trend to regionalism. Ethnic groups, social persuasions, "cults" of many different kinds, all take advantage of any available platform to publicize their identity. A recent issue of the magazine *Daedalus* is entitled "The End of Consensus" to symbolize the fact that practically all the systems of beliefs and tastes which had cemented Western societies until now are in the process of breaking apart—thereby generating a kind of global anarchy. Folklores, local historical societies and museums are in fashion—witnesses to the fact that regional loyalty is more natural to humankind, both biologically and socially, than is either rootless drifting or belonging to a nation-state. Every ethnic group envies the black people of America for having as one of its members the author of the international best-seller *Roots.* There is still much intellectual talk about the global village and the melting pot, but the topic of *Roots* has more emotional meaning. Vast movements of population have occurred throughout history, but the migrants eventually settled down to become the local peasantry, bourgeoisie and aristocracy. The Saxons and Northmen who roamed through western Europe in the Dark Ages and who moved into England from French Normandy under the leadership of William the Conqueror became in a few centuries the very proper people known as Anglo-Saxons and now as WASPs. Mass population movements still occur, but they do not last long. Even in the United States, the much-vaunted mobility eventually evolves into a local, regionalist culture. It does not take long for Scandinavians to become conservative Midwesterners, for Italians and Mexicans to become loyal Californians, and for Puerto Ricans to become sophisticated New Yorkers. Poets write about the open road, but few people elect hobo life. Mobile homes can be seen in huge numbers all over the United States but most of them are anchored in trailer parks, not moving on highways.

Regionalism in the United States has the special interest that its historical determinants are of recent origin. In any given area, the percentage of natives is almost always small, but dedication to the spirit of the place is likely to be the highest among the newcomers. The people most eager to save the Vermont or Oregon countryside have usually moved into these states from New York, New Jersey, California or any area where they feel that the countryside has been spoiled. Regionalism used to be the expression of natural forces and historical accidents, but this is less the case now than it used to be. In most cases, choice rather than the place of birth makes for local patriotism. Greater geographic and economic mobility has increased the latitude in the selection of a place of residence and it can be expected that conscious choices will increase the desire of people, indeed their eagerness, for a more local management of their communities and of their environment.

Regionalism was highly developed on the American continent among the original Indian populations and during the early phase of European occupation. The sense of the region, however, was much weaker or nonexistent among the immense numbers of people who immigrated during the nineteenth and twentieth centuries. Many of these people had acquired their cultural identity in Europe by long periods of settlement in one given region, but the immigrants who became the new Americans derived their dominant cultural traits from the conscious choices they had made to abandon their past and adopt certain lifestyles and social institutions they hoped to find in America. Whether originating from an isolated Swedish farm or a crowded Polish ghetto, the new immigrants longed for a democratic way of life in the land of unlimited opportunities. They had decided a priori to become "Americans" and this was more important in determining their tastes and behavior than the characteristics of the region from which they originated or in which they happened to settle, often by chance and temporarily. As Talleyrand wrote in one of his letters during his several months' sojourn in the United States, "Tout homme qui choisit ici sa patrie n'est-il pas d'avance un Americain?" He had noticed that most Europeans who had elected to settle in the United States immediately adopted attitudes and ways of life that appeared to him those of a frontier situation rather than those of European civilization.

In his famous essay, "The significance of the frontier in American History," first published in 1893, the American historian Frederick Jackson Turner had formulated the theory that the cultural traits of the people in the United States and the characteristics of their institutions had been largely shaped by the experiences of the settlers as they moved from the Atlantic to the Pacific Coast. Turner used the word "frontier" not in its usual sense of a limit of demarcation between geographical entities but rather to designate an ill-defined area constantly changing with the movements of the settlers and in which economic practices, administrative structures and ways of life were unsettled and rapidly evolving. Historians now doubt that the frontier experience, as defined by Turner, played as formative a role in the American psyche as he claimed; but it is certain that the American imagination has been deeply imprinted by the frontier *myth* and that geographic and social mobility, as well as a pioneer spirit, have been generally accepted as parts of the American ethos and temperament.

Now that the continent is fully settled, regionalism seems to be gaining ground once more in the United States. Turner himself had anticipated that such a change would occur after the end of the frontier experience. In his essay, "The Significance of sections in American History," published in 1932, he asserted that "National action will be forced to adjust itself to conflicting sectional interests." His essay dealt chiefly with the economic and political differences between the various regions (for which Turner used the word "sections"), but there is no doubt that regional differences affect many other aspects of life, from the most personal tastes and cultural attitudes to the most technological practices.

The development of cultural regionalism in the United States is likely to be accelerated by the fact that more and more people can now afford to select the region where they settle. Furthermore, many people are not only place seekers, but place makers, interested in the socioeconomic and cultural potentialities of the place where they have elected to settle. Some people long to relax on a Florida beach, others to experience the exciting Far West atmosphere, and still others would rather saw firewood in New Hampshire. There is furthermore some likelihood that population mobility will decrease in the future, not because of economic difficulties but because of changes in behavioral patterns. The eagerness for mobility may be toned down by the belief that civilization does not depend on constant movement, and is indeed more likely to flourish when populations are anchored on the earth.

It is easy to understand why physical characteristics make life-styles in Florida or California different from those in the Middle West and New England. But other reasons of a less obvious nature are likely to increase the range of regional differences in the near future. The scarcity and cost of certain fossil fuels may compel us to use other energy sources which are renewable and not polluting, such as solar radiation, the wind, the biomass, the tides and waves. For a multiplicity of reasons, all these potential sources of renewable energy tend to be localized, a fact which will probably bring about changes in the geographic distribution of various industries. For example, harnessing solar radiation economically with reflectors might bring about the emergence of new industrial centers in the Southwest, whereas the use on a large scale of the biomass as source of energy might favor industrialization of the Southeast and Northwest.

Another factor which has made for uniformity in the past has been the concentration of various types of agriculture and of animal husbandry in specialized parts of the world. This trend, however, may not continue. Complete dependence on external sources of food is fraught with dangers even in the United States. For example, there is a possibility that California, Texas and other food-producing states may become so populated that they will consume most of the food they produce and therefore will have little to export to the Northeastern states. Furthermore, shipment of certain kinds of foods over long distances may eventually become undependable because of labor conflicts and prohibitively expensive because of energy costs. These prospects point to the possibility that it will become desirable once more, at some time in the future, to produce certain kinds of crops and livestock which are adapted to regions where agriculture has been all but abandoned. A partial degree of independence with regard to food production is in fact coming to be considered a matter of national security in most parts of the world.

Thus, human beings can satisfy their essential needs in many different ways depending upon the resources available in the place where they live and the circumstances of the time. While animals and plants respond to the challenges of their environment almost exclusively by adaptive changes in their biological nature, human beings usually respond to such challenges by changing their social habits and also their environments so as to make these better adapted not only to their biological needs, but also to their fantasies.

Each type of civilization is characterized by its own ways of life and of shaping its natural environment. In the past, for example, there developed empirically and spontaneously all over the world regional types of shelters determined by climatic conditions, local resources and social habits. This "architecture without architects" was picturesque and contributed much to the genius of the place. It was largely replaced during the twentieth century by the more uniform international architecture when the low cost of petroleum and natural gas made heating and air-conditioning economically possible. The ever-increasing cost of these fossil fuels and of other forms of energy is once more leading to new forms of architecture and of planning better adapted to local natural conditions and to modern ways of life.

The range of these social adaptive changes is so wide that no book, however large, could give a comprehensive view of them. I shall therefore illustrate them with only two contrasting examples of contemporary human settlements in which extremely high population density and shortage of space created special and difficult problems of environmental management. One example is the Netherlands, a nation where the problems of accommodating a large population has been solved by creating the most technologically developed horizontal countryside. The other is Manhattan (originally called New Amsterdam!) where the same problem has resulted in the creation of the world's most vertical city.

THE NETHERLANDS, HORIZONTAL COUNTRY CREATED BY HUMANKIND

The planners of Imperial Rome certainly did not consider the small area of northwestern Europe now occupied by the kingdom of the Netherlands as a promising place for the development of civilization. Few parts of the world have been as poorly endowed by nature with regard to climate, soil fertility and other resources. Julius Caesar did take an interest in the Low Countries, and Roman legions occupied much of them at the beginning of the Christian era, but only for military reasons. The Low Countries* were in a strategic position for the defense of the Empire's western provinces as they straddled the route by which the barbarians from the German forests could penetrate the rich agricultural lands of Gaul and southern Europe.

To Mediterranean eyes, the Low Countries must have appeared most undesirable at that time, just a desolate area of heath, fetid swamps and dank woodlands, unsuited to agriculture and with no mineral resources. The North Sea climate is commonly at its worst in this region—cold, humid, foggy and with winds that are almost constant and at times extremely violent. The country was nevertheless extensively and intensively developed early during the historical period to such an extent that, even before Spanish occupation, much land had to be protected by a complex system of dikes, levees and pumps against inundation by the waters both of the North Sea and of the rivers.

Peat bogs abound throughout the Netherlands and much of the land is below sea level. The greatest part of the remainder of it is made

* The phrase "Low Countries" applies not only to the kingdom of the Netherlands, but also to parts of Belgium. The name Holland refers specifically to a particular province in the southwestern part of the Netherlands which, because of its hegemony among the United Provinces after the successful revolt of the Dutch against Spain, came to be applied loosely to the whole of the Dutch Republic and later to the kingdom of the Netherlands.

up of low sandy plains with occasional hill ridges rarely exceeding 45 meters in elevation. In practically all parts of the country, the soil can be used for agriculture only after elaborate techniques of reclamation and of fertilization. Even though the Netherlands had become one of the richest countries in the world at the time of Spanish occupation in the sixteenth and seventeenth centuries, with an extremely productive agriculture, there was some justification for the remark attributed to the Duke of Alba who had tried to overcome the Dutch rebellion against Spain, "Holland is as near hell as possible." There is much truth, however, in another remark attributed to a Frenchman, "God created the world, but the Dutch *made* Holland." The word "made" is appropriate because the Netherlands provide the most spectacular example of humankind's ability to transform the surface of the earth and to create out of it artificial environments in which animals and plants can prosper and civilization can develop—a process that I have called elsewhere "the wooing of earth" by humankind, or creative symbiosis between earth and humankind.

My purpose in this chapter is to give a brief account of the steps by which the Dutch have succeeded in creating extraordinary sceneries and a prodigious civilization on what sociologists have facetiously called a bad piece of real estate. A few details of ancient history and even prehistory will show that the unique features of urban settlements in the Netherlands have progressively emerged from very primitive techniques of land management.

Uninviting as the natural conditions appear to be for human life in the Low Countries, *Homo sapiens* had nevertheless become established along the coast more than 10,000 years ago. Farming and the keeping of cattle were already carried out during Neolithic times and the early settlers burned over the heaths to encourage new, tender growth for their flocks of animals. This practice and overgrazing progressively destroyed much of the vegetation and exposed vast tracts of the underlying sand. Under the influence of the wind, much of that sand was pushed inland where it created dunes in many areas.

Long before the Roman era, people originating from Germany and other parts of eastern Europe had built fairly extensive settlements on salt marshes barely one-third meter above sea level, especially in northern areas which are now the provinces of Groningen and Northern Friesland. As these settlements were dangerously liable to flooding, they were repeatedly heightened by layers of turf sod and of clay. Such refuge mounts, commonly designated *terpens,* became more and more numerous. They were at first just spacious enough to protect a single farmstead or hamlet, or to serve as refuge for livestock during times of flood. As terpens progressively increased in size and height, they provided a pattern for the future development of most Dutch towns. The population must have increased considerably even before the Christian era as indicated by the fact that the single Germanic Batavi tribe, which had settled in the northern part of the Netherlands, could supply 10,000 men as auxiliaries to Julius Caesar's Roman legions.

Roman rule vanished completely in the Netherlands around 350 A.D., leaving little trace except for a few settlements and some faint evidence of the grid pattern in the countryside. During the troubled centuries that followed the Roman withdrawal, people from eastern Europe—Frisians, Saxons and Franks—established themselves all over the Netherlands; the Dutch population largely stems from them. Like the rest of Europe, the Netherlands suffered from invasions by Vikings and other Nordic people during the Dark Ages.

Relative peace and order began to prevail after the end of the Nordic invasions and this permitted rapid agricultural and economic development. Since much of the land is protected from the North Sea only, and ineffectively, by the sand dunes, strict control of the water from the sea and from the rivers was essential almost everywhere. The process began when the early settlers learned to build terpens as refuges against storm tides and river floods. The technique was sufficiently developed in the eighth or ninth century for cities such as Leiden and Middleburg to be established along the western coast on artificial mounds 100 or so meters in diameter and some 15 meters high—real hills in this very flat country.

In most places, the ground is so wet that the only way to put up a new building is to prepare a special site for it. While onerous, this necessity has had the advantage of preventing haphazard growth and urban sprawl. In contrast to the disorderly urban development all over the United States and in much of Europe, the Netherlands offer a crisp break between town and country and this makes possible ready access from urban settlements to bucolic scenes. A ten-minute drive from almost any town leads to a place beside a river or canal where one can sit on the grass, admire wild flowers, watch many more kinds of birds than one is likely to see in other Western, industrialized countries.

Protection against water depends largely upon dikes. The first substantial ones were built in the eighth or ninth century. Most parts of the coastal belt had thus been protected against the sea and the rivers by the end of the thirteenth-century. Time and time again, however, villages were devastated and whole areas of agricultural land ruined as a consequence of flooding from the rivers or after the bursting of dikes. On the night of 18–19 November 1421, for example, a great storm tide backed by a fierce westerly wind attacked and breached the tides at Brock and destroyed much land north and south. Early accounts speak of seventy-two villages destroyed and 100,000 persons drowned. Modern studies suggest that these figures may have been exaggerated but the destruction was so great nevertheless that many people, including the nobility, had to go begging and that their plundering raids made the region unsafe for several years. This is but one example among the countless disasters caused by wind and water throughout the history of the Netherlands. The disastrous floods of January 1916 and of February 1963, to which I shall return later, show that the threat is always present and especially dangerous in the Holland and Zeeland areas. Throughout history, new dikes were built constantly and rapidly to replace the ones that had been destroyed, an enterprise that became and has remained a national commitment. The motto of the arms of the province of Zeeland, *Luctor et Emergo* (I strive and I arise), symbolizes the spirit of the Netherlands as a whole.

As already mentioned, the early settlers found little good fertile soil. Agricultural land was created at first from peat bogs, especially until the nineteenth century when peat was the chief fuel used in the Netherlands. A spectacular feat of such land reclamation took place from an extensive peat bog in the northeastern part of the country. Beginning in the seventeenth century, the peat was dug up, then the sterile sandy subsoil was progressively changed into rich farmland by the addition of manure and of artificial fertilizers.

However, the most typical and picturesque type of artificial farmland is in the polders created from completely submerged areas. Lakes, whether of salt or fresh water, are pumped dry after having been surrounded by a dike and a wide canal. The canal is dug first and the soil removed from it is used to build the dike around the area to be converted into a polder. The water is then pumped into the canal which carries it to a river or to the sea. Polders are usually crisscrossed by small ditches which keep them well drained. The layout of their meadows or crops, cut into regular strips by the ditches, calls to mind a Mondrian painting, or more prosaically the gridiron pattern of a city the streets of which would have no traffic. Standing below in the polder itself provides the sight, extra-ordinary in the strong etymological sense of the word, of boats in the encircling canal silhouetted against the sky as they move above the land.

Most of the tremendous tasks involved in the construction, maintenance and repair of the dikes, the development of levees along the canals and rivers and the creation of land from peat bogs and in the polders were originally carried out with simple tools. Spades and mattocks sufficed for most construction work; heavy pile drivers, operated by teams of thirty to forty men, were used to build levees and dams. Large wicker baskets, sleds and ox-drawn carts served for the transport of material, such as clay for the dikes, or timber piles and mattresses of willow wands to reinforce the water defenses. The wheelbarrow was not introduced until fairly late.

Pumping out the water was naturally the crucial part of the reclamation, and this used to be done by thousands of windmills, the thatched roofs and moving sails of which became a picturesque and endearing feature of the Dutch landscape. The use of windmills, however, is of more recent origin than commonly believed. For many centuries the only methods for removing excess water from the polder were by gravity drainage, the counterbalanced bailing bucket, and the horse- or man-powered scoop wheel. The first two records of wind-driven water mills date from 1408 and 1414. These first windmills were small and inefficient. Larger and more efficient ones, working on the principle of the Archemedean screw and able to lift water for two to four meters, did not become common until after the beginning of the seventeenth century. They are the water mills made familiar by pictures of the Dutch scenery.

Needless to say, the control of water and the creation of land demanded tremendous efforts that exceeded the abilities of a single family or a small group of families. The building of houses in areas artificially protected from water required much sand carrying and pile driving, and implied not only communal participation but also a high level of social organization and discipline. In the Middle Ages, severe laws were enforced to help win the battle against water. "No feuds were allowed once the dikes needed repairs," writes Dr. J. van Deen, the historian of Dutch reclamation. Breaking the dike peace could mean being sentenced to death. "In some parts, any man refusing to do his share could be buried alive in the breach with a pole stuck through his body. The people who lived further inland had to come and work at the dike. 'Dike or depart' was the old saying. Any man who was unable to repair the breach in his own section of the dike had to put his spade in the dike and leave it there. This was the sign by which he gave his farm to any man who pulled the spade out of the dike, and who felt powerful enough to close the breach. This was the *Law of the Spade."* Eventually there arose administrative structures which derived their authorities from their services to the common good. The personal and social disciplines which were essential conditions for water control must account in large part for the complementary spirits of independence and of democratic tradition which have contributed so much to the Dutch ways of life and to the greatness of the Netherlands.

Despite incessant struggle against the elements and in spite of incredibly complex political difficulties both at home and abroad, most semiindependent parts of the Low Countries had achieved great prosperity at the time they became part of the Spanish Empire. A measure of their power is that they eventually achieved freedom from Spain, which was then the greatest and wealthiest military power in Europe. The Dutch even established six new universities in different parts of the country during and immediately after the Spanish War.

The phenomenal achievements of the Netherlands are the more striking when it is realized that, at the time of the Spanish wars, the population was only of the order of three million people grouped in more or less independent municipalities within seven provinces. These became the "United Provinces" for the purpose of eliminating Spanish domination but they were otherwise in constant conflict about religious, economic and almost any other matter. The religious aspects of these rivalries are especially entertaining because they involved not only Catholics against Protestants, Calvinists against Lutherans, but even bitter conflicts of doctrine between different groups of Calvinists.

The Netherlands thus had a form of government that was part oligarchic, part republican, unique in its diversity and for this reason seemingly far less effective than the surrounding European monarchies, from both the political and administrative points of view. The success of the Netherlands in the face of so many internal and external difficulties provides evidence that the local management of affairs is usually more effective than national or global management.

The sixteenth and seventeenth centuries have been called the Golden Age of the Netherlands. The western part of the country, i.e., Holland proper, and Amsterdam in particular, were then the envy of western Europe. The three million people of the Seven United Provinces which had been so poorly endowed by nature had not only triumphed over

Spain; they dominated the maritime trade of the world and managed much of its finance. Their merchant marine was ten times greater than that of France, three times greater than that of England; indeed it was greater than those of England, Spain and France combined. The commercial enterprise of the Dutch carried their merchant vessels from Recife to Nagasaki and from Archangel to the Cape of Good Hope.

The Netherlands were also leading in practically all fields of culture. Rembrandt the painter, Spinoza the philosopher and Dr. Nicholas Tulp all lived in Amsterdam at the same time. Dr. Tulp is of special interest to me for two different but related reasons. He taught the then unorthodox doctrine that the chief role of the physician is not to administer drugs but to help the patient mobilize the natural defense mechanisms which are essential to the control of disease. He was painted on several occasions by Rembrandt, who also made of him an etching that I regard as the most perceptive portrait of the ideal physician. In nearby Delft, there also lived at that time Anton van Leeuwenhoek (1632–1723), the first scientist to develop a microscope that enabled him to see and follow the activities of bacteria and other microbes. Several illustrious foreigners also lived in the Netherlands during the Golden Age. Tsar Peter the Great of Russia had come to study shipbuilding. The French philosopher René Descartes found Amsterdam the best place in which to think freely in the midst of great comfort. As he wrote in one of his letters, "What other country could one choose where all the conveniences of life and all the exotic things one could desire are to be found as readily? Where else could one enjoy a freedom so complete?" Amsterdam was then indeed *the* city of Europe.

The agricultural productivity, as well as the expansion of industry, commerce and overseas trade had all been achieved despite extreme paucity in natural resources and against great handicaps created by natural and political conditions. They were the results of human efforts, intelligence and daring symbolized in two paintings by Franz Hals. These paintings capture contrasting aspects of the Dutch character, both of which contributed to the success of the Netherlands at that time. One conveys the Puritan sobriety of the Governors of the Old Men's Home at Haarlem and epitomizes the industriousness which then prevailed in all Dutch towns. The other shows the roistering cavaliers of the St. Jorisdoelem—the swashbuckling adventurers who carried the Dutch flag around the world. The capacity for assiduous toil and the daring spirit can also both be recognized in the urban architecture of that time—the staid simplicity of most churches and almhouses contrasting with the splendor and the self-assurance of town halls and merchant dwellings.

The qualities of Dutch life are evoked by Albert Camus in his novel *La Chute* (The Fall). "You take these good people for a tribe of syndics and merchants counting their gold crowns with their chance of eternal life, whose only lyricism consists in, occasionally, without doffing their broad-rimmed hats, taking anatomy lessons. You are wrong. . . . Holland is a dream, monsieur, a dream of gold and smoke—smokier by day, more gilded by night. . . . They have gone thousands of miles away, toward Java, the distant isle. . . . Holland is not only the Europe of merchants but also of the sea, the sea that leads to Cipango, and to those islands where men die mad and happy. I like these people swarming on the sidewalks, wedged into a little space of houses and canals hemmed in by fogs, cold lands, and the sea steaming like a wet watch. I like them because they are double. *They are here and elsewhere* [italics mine]." They are "here" because, since the beginning of time, the Dutch have lived in environments they have created, almost entirely by themselves, out of nature; they are in consequence integral parts of these environments. But they are also "elsewhere" for two different reasons. On the one hand they became early part of the whole world not only because their ships made them the greatest international carriers of merchandise but also because their technical proficiency in many fields gave them the opportunity to work in the most unexpected places—draining the English fens and the Russian marshes, popularizing and selling almost everywhere tulip and other bulbs originally derived from the Orient, converting the Philips industrial plant in Eindhoven into a center for electronic research with branches in many parts of the world. On the other hand, the Dutch have long lived in such crowded environments that they can remain sane—as they have most of the time—only by spending much of their lives within their own thoughts, aware of the external world, of course, but concerned at least as much with their own private, internal world.

The wind-driven water-pumping mills which made possible the great reclamation schemes of the seventeenth century and which were so typical of the Dutch landscape during the Golden Age began to be replaced by steam pumps at the end of the eighteenth century. But it was the introduction of the electric pump at the beginning of our own century which made possible the draining of the Zuider Zee, one of the greatest technological achievements of modern times.

The Zuider Zee used to be a very shallow body of water, silted with rich alluvions from the several rivers that empty themselves into the North Sea. Plans to drain it and convert it into farmland had been formulated in the seventeenth century but were then beyond the possibilities of technology. More precise schemes were put forward during the nineteenth century and the one submitted by the engineer Dr. Cornelis Lely was adopted in principle by the Dutch parliament in 1901. A great stormtide in 1916 and the food shortages caused by the First World War finally led to passage of the act for the closure of the Zuider Zee in 1918. The scheme was to throw a dam across the whole mouth of the Zuider Zee in order to cut it off completely from the North Sea and then to proceed gradually with the draining operations. The scheme had three objectives: to reclaim some 555,000 hectares of dry land (i.e., approximately one tenth of the agricultural area of the Netherlands!) in several great polders; to reduce the coastline by 300 kilometers; and to provide a freshwater reservoir of 296,000 hectares. One of the polders (Wieringermeer) was completely drained in 1930, two years before the dam was completed.

Work on the project was begun in 1923 and the last gap on the dam was sealed on May 28, 1932. A monument on this spot reads: "Een Volk dat leeft bouwt aan zijn toekomst" (a living nation builds for its future). Faced with interlocking blocks of Rhineland basalt, the dike crown rises 7.5 meters above mean sea level. It is 32 kilometers long and carries a wide motorway that links Friesland to North Holland. It has locks for ships to pass in and out. There are sluices and a large pumping station which regulates the level of the water inside the dam. What was once the Zuider Zee now consists of four polders and a freshwater lake, the IJselmeer. The latter supplies drinking water to the human settlements on the polders and also water for irrigation in periods of drought.

It was taken for granted in the 1930s that all the reclaimed land would be used for agriculture; two of the Zuider Zee polders are indeed devoted to farming. In the Netherlands as in many other countries, however, economic prosperity is taking people out of agriculture, changing their habits, and increasing their expectations. Fewer and fewer people want to live in the small hamlets typical of the old polders. Motorcycles and cars make it possible for them to live in larger places which provide more services and attractions. Furthermore, now that the Netherlands are part of the Common Market, they can obtain certain crops more cheaply from Italy and France than from national agriculture.

Fewer and fewer Dutch people are engaged in farming and more and more of the reclaimed land is being used for new towns, industry and various forms of recreation.

The reclaimed Zuider Zee remains in any case an almost miraculous example of the transformation of the surface of the earth by human ingenuity and effort. I visited the area in the fall of 1969, on a rather stormy day. Listening to the violent wind and to the waves lashing on the dam, it was difficult for me to imagine that, only two decades earlier, there was only salt water inside the dam with fishing boats and screaming gulls where I now saw substantial farms and handsome villages, with religious buildings around the village square. (As I recall, Protestant, Catholic, and Jewish congregations had to be represented in each village.) A cup of hot rich chocolate in one of the small taverns made the experience even more like a dream.

For many decades now, some two percent of the national income of the Netherlands has been spent on dredging, draining and reclaiming for the creation and protection of dry areas. Much land has been thereby made available for agriculture and other human activities, including the enjoyment of "nature." But the environment is called "nature" only for lack of a more accurate word, since most of the surface of the earth has been profoundly transformed by human intervention. Two figures will suffice to illustrate the magnitude of this transformation. In 1840, the salt coastline of the Netherlands was 1979 kilometers long. It will be reduced to 676 kilometers after completion of the system of dams, sluices, storm barrages, and strengthened dikes which are being completed in the 1980s along the vulnerable coastline of the Rhine-Maas-Scheldt Delta. "The Dutch have truly made Holland" by transforming nature through the bold use of technology, and what is even more remarkable is that the process of transformation of the bodies of water and of land continues to go on at an accelerated pace in many parts of the country.

The recent feats of land reclamation are technological miracles of the twentieth century but their very complexity makes certain human settlements even more vulnerable than in the past to violent storms and also to accidents of human origin. For this reason, watch is kept constantly along the dikes and weather forecasts are announced twice daily in winter.

On January 31, 1953, the wind became increasingly violent during the day and at 11:00 P.M. the radio warned that the following day would bring "severe northwesterly gales . . . unsettled weather, showers, hail and snow." On February 1, the northwest gales increased the tides, which are in any case normally highest at that time of the year, and pushed a huge mass of water through the North Sea. The sea rose higher than it ever had in human memory, washing over the dikes in some protected areas and smashing through them in many places. Close to 1280 kilometers of dikes were destroyed. Miles of railroad track were washed off the top of a dike 10.4 meters above sea level. The sea water came to cover more than 400,000 acres of farmland and villages; almost 2000 persons lost their lives; tens of thousands of animals were drowned; similar numbers of buildings were destroyed or damaged; the soil was poisoned by the sea salt. The dikes were repaired within a year, at the stupendous cost of six percent of the national budget, but it took seven years before the Dutch agricultural production fully recovered from the February 1953 disaster.

After the 1953 catastrophe, a Delta commission was organized to explore the possibility of sealing off the Scheldt and Rhine estuaries and of shortening the southern shoreline. The Delta project was adopted in 1957 and when completed, during the 1980s, should greatly enhance the safety of the southwestern Netherlands. Its main purpose is not, as was the case for the Zuider Zee, to reclaim new land but to increase the country's freshwater resources, to check the penetration of salt into existing agricultural land and to make the islands in the Delta more accessible. These islands could then be used for industry, more intensive agriculture, or as residential and recreation areas much needed in the southern Netherlands. The Delta project was altered by an act of Parliament in 1976 under the pressure of environmentalists who pointed out that the initial plan would irreversibly change the ecological character of the estuary. The controversies on the merits of the new, much more expensive plan, continue and thus make it clear that the Dutch are still in the process of making Holland, still hoping that the eastern Scheldt will be closed to the ravages of the North Sea in 1985.

Another spectacular project still in the process of development is along the thirty-five kilometers of waterfront that separate Rotterdam from the North Sea. Although Rotterdam was practically destroyed by the German air raid of May 1940, its harbor was rapidly rebuilt after the war. More importantly, it has been modernized and enlarged to such an extent that it is now the largest, busiest and most modern harbor in the world, thanks to its location close to the highly industrialized and densely populated areas of western Europe, with some 250 million people within a radius of five hundred kilometers. Everything seems to be the biggest and most modern in the Rotterdam area: the world's largest container loading and unloading facilities; one of the highest capacity grain terminals; and accommodations for the largest tankers.

The river Nieuwe Mass which crosses Rotterdam from end to end now reaches the sea through an artificial channel called the Nieuwe Waterweg (New Waterway) which was first dug between 1866 and 1872. Gigantic dredging operations have progressively deepened the New Waterway to a depth of 23 meters over a length of 11 kilometers, and the process of extension continues. The docks are equipped to deal rapidly with the largest ships, including the most monstrous oil tankers.

The new harbor section is properly called Europort because it serves as a European processing center for handling and redistributing grain and oil as well as for the production and distribution of countless chemicals. Oil storage tanks, refineries, chemical plants extend for many kilometers out of Rotterdam. But the most impressive aspect of the enterprise is the extent to which the city, as well as the harbor and industrial facilities of Europort are new, artificial—a purely human creation which has completely transformed the natural environment.

The phenomenal density of population and intensity of industrial growth in the western Netherlands would have resulted in massive environmental degradation if it were not for the local peculiarities of urban development. With some six million inhabitants, this area is the most densely populated in the world—some 1000 persons per square kilometer. If the United States were as densely populated as the Netherlands, the total American population would be as large as that of the whole world today—more than four billion people. In the Netherlands, furthermore, six out of ten people live on land which is below sea level. At the present time thirty-seven percent of the Dutch people dwell on five percent of the country's area and it is all but certain that the population will continue to increase. Most people, furthermore, live in a horseshoe-shaped belt of towns, cities and suburbs called the Randstad (Ring City) located in the Holland province. The base of the horseshoe is on the North Sea dunes and its open end faces southeast. It is about 48 kilometers across, would be 177 kilometers long if straightened out and runs from Dordrecht through Rotterdam, Delft, The Hague, Leiden, Haarlem to Utrecht. Its population density is four times higher than the average for the country as a whole.

Despite this extreme level of industrialization, urbanization and crowding, most Dutch people enjoy excellent health, with a long expectancy of life and a low crime rate. There are still small stretches of open space between most of the cities and towns of the Randstad but the most remarkable and uniquely Dutch aspect of this conurbation is that its core retains a rural appearance.

The horseshoe area between the towns and the North Sea—or rather between the mounds on which the towns are built and the dikes and dunes that separate them from the North Sea—is rich agricultural land with well-tended farms. Cows, ducks, swans, herons, lapwings and of course, the water of the canals and rivers, can be seen everywhere. As mentioned earlier, this happy situation is due to the fact that human settlements have had to be built on mounds and therefore cannot sprawl as is the case in the industrialized areas of other countries. In consequence, cities, towns and even villages have to be sharply separated from the country where grass and flowers are stirring with the wind.

The wind itself contributes to the diversity of landscapes and waterscapes. It blows almost constantly, brushing over the dunes and across towns and meadows, bringing low clouds and rains from the Atlantic in all seasons. It carries now and then the summer warmth or the freezing cold from central Europe; it bends trees and also turns the few surviving working windmills. Seen from a low-flying airplane, this tiny region which is so tight and tense, busy and booming in the human settlements of the Randstad, appears, nevertheless, more comfortable than other countries better endowed by nature. Its agricultural core of meadows, cultivated fields and water might well serve as a model for social organization and land management in rich urbanized industrial countries.

Recent developments in the city of Amsterdam will illustrate the extent to which the Randstad landscape and waterscape are completely managed. The Amsterdam Forest Park, called "The Bos," or more colloquially "the Woods," was created southwest of the city in 1934 by sacrificing three polders 4 meters below sea level. An extensive network of drainage pipes had to be laid to maintain a water level compatible with the development of the roots of the trees to be planted. Lakes created by the drainage are used for swimming and canoeing. The shape of the steep polder dikes were softened into gentler slopes and an artificial hill was created. All sorts of trees common to northwestern Europe were planted over one thousand of the total 2230 acres. The Bos woods, now almost half a century old, consists of long wooded slopes, lakes and canals where people walk, ride bicycles, fish or row their canoes. Icelandic ponies can be seen ambling through the trees;

but this bucolic picture acquires a less idyllic aspect when more than 100,000 people come to the woods on summer weekends.

Paradoxically, the most disturbing aspects of the Netherlands' future may originate, not from shortages in resources or economic difficulties, but from the very successes of the Dutch people in shaping their country and from the consequences of these successes for the safety and the quality of future Dutch life.

For example, the growth of Amsterdam has required the development of huge dormitory cities. On the west side, the creation of Slotervaart, Slotermeer, Osdorp and Geuzenveld involved raising the subsoil level by at least 1.8 meters. This was done by sacrificing and excavating the huge Sloterplas polder over an area approximately 1.6 kilometers long and 0.4 kilometers wide, to a depth of 27.4 meters. The rich polder topsoil was used to create parks, lawns and recreation land; the 19.9 million cubic meters of sand that was excavated from beneath served to prepare the building site. The excavation site itself was turned into a lake.

All these new housing developments differ from the traditional cities and towns of the Netherlands, and from central Amsterdam in particular, in offering large open areas either of land or of water. This change, however, is considered undesirable by many people who regard the dormitory towns as too orderly, too inflexible, with no pleasurable surprises. The absence of Dutch coziness suggests that even though the buildings and parks may be of high technical quality, they are the kinds of settlements that municipal authorities design for "other people" but in which they themselves rarely elect to live.

The city of Rotterdam also symbolizes both the triumphs of technology and their threats to the quality of life. Rotterdam can legitimately boast of having the largest and most modern harbor facilities of Europe and perhaps of the world, but this technologic and economic success has been achieved at the cost of many other values. The deeper the New Waterway is dredged, the larger the amounts of salt and toxic substances that enter from the sea and that contaminate not only the drinking water but also the soil of the agricultural Westland. This trend will continue even if, as is claimed, the Delta plan increases the volume of Rhine water that will flow through the New Waterway.

Air pollution from oil refineries and from chemical industries is constantly increasing. The oil spillage from the tankers is contaminating most North Sea beaches. Tar contamination is so common that people make it a practice to keep a bottle of gasoline with a rag at the entrance of their house to clean the soles of shoes before walking in. Many Dutch people are beginning to doubt that there is a real economic or other justification in one more oil tank, one more distillery, or one more plant for the manufacture of plastics. They are asking themselves whether they really want more growth or whether they are continuing to build, not out of necessity but simply because they know so well how to build.

Perhaps more important in the long run, however, is the threat to the coziness of the environment in the Netherlands, and particularly to the intimate relationship between people and things that used to be such an appealing aspect of Dutch life. Will the exciting environments now being created by modern technology ever have the appeal of the exquisite atmospheres conveyed in the paintings by Rembrandt, Jan Vermeer, Pieter de Hooch and the other masters of the past—atmospheres which made the rest of the world so envious of Dutch life during the Golden Age of the Netherlands?

MANHATTAN, THE VERTICAL CITY

Just as in the case of the Netherlands at the beginning of the Christian era, Manhattan Island and its surrounding waters did not appear of great promise to the navigators and explorers of the sixteenth and early seventeenth centuries. The first Europeans to see Manhattan Island were a group of French and Italian sailors under the command of Giovanni da Verrazano, an Italian navigator sailing in the interest of Francis I, king of France, to find a westerly passage to the Indies. In 1524, Verrazano navigated his ship, *La Dauphine,* along the eastern coast of North America and found what he termed "a very agreeable situation located between two small prominent hills [the Narrows] in the midst of which flows a very great river, very deep within the mouth." Verrazano sailed a small boat into the Upper Bay of the Hudson which he described as "a very beautiful lake" but a gale arose and forced him back to *La Dauphine,* where he weighed anchor to travel further east. A year later, in May 1525, Estevan Gomez, a Portuguese pilot in the service of Emperor Charles V of Spain, also seems to have entered New York Bay. In 1570, Jehan Cossin, a French navigator from Dieppe, drew maps which prove that he explored the outer and inner bays of New York City. None of these navigators seem to have returned to this area, as if they did not consider it of much importance.

In 1609, Henry Hudson, an English adventurer employed by the Dutch East India Company, also tried to find a western route to the Indies. He passed through the Narrows but unlike Verrazano he took his ship, the *Half Moon,* as far as Albany on the river that now bears his name. On his return to Amsterdam, Hudson reported to his Dutch employers that the natives he had seen possessed beautiful furs. This caused several Amsterdam merchants to send another ship to the Hudson River in 1610. Finally, other Dutch ships arrived on Manhattan Island itself to establish a trading post. This led to the formation of the Dutch West India Company and in 1624 the ship, *Nieu Nederlandt,* set sail from Amsterdam with thirty families and arrived in early May at the mouth of the Hudson River. Most colonists, however, sailed up the Hudson and settled near the site of the present city of Albany. The first stable settlement on Manhattan Island did not begin until May 1626. A few months later, Peter Minuit, who was then director-general of the Dutch colony, bought the island from the Indians for trinkets valued at sixty guilders—a sum said to be the equivalent of twenty-four dollars. The Dutch settlement, christened New Amsterdam, began to prosper only under the administration of the director-general Peter Stuyvesant, who arrived in May 1647 and ruled it until he had to surrender it to the English in 1664. Toward the end of 1664, the name New Amsterdam was changed to New York even though the population was still chiefly Dutch.

At the beginning of the Revolutionary War, New York was a small provincial town of some 12,000 inhabitants, most of whom were Dutch or English; the Dutch language was still being used by a large percentage of the population. Trade with the Indians and with Europe, chiefly about furs, was the mainspring of prosperity. New York was occupied by the British during most of the Revolutionary War but, as is well known, it was at Fraunces Tavern in Manhattan that George Washington took farewell of his officers in 1783.

New York became the center of the Federal government in 1785 and four years later on April 30, 1789, George Washington was inaugurated as President of the United States, taking the oath of office in Federal Hall on Wall Street. Difficult as it is to believe, the first Federal Census, taken in 1790, showed a population of only 33,131 people. New York was the capital of the United States until 1790, then became the capital of New York State until 1797 when the state capital was moved to Albany.

Manhattan, which had long owed what little prosperity it had to farming and the fur trade with Europe, acquired a new importance when, on November 4, 1825, Governor De Witt Clinton formally opened the Erie Canal, a waterway cut from the Hudson River to the Great Lakes. From then on, the agricultural and industrial production of the Middle West had easy access to New York and this increased the importance of the harbor. In 1838 two ships, the steamers *Sirius* and *Great Western,* crossed the Atlantic Ocean by steam power alone, thus establishing rapid communications between the European and American continents, and creating thereby the commercial supremacy of Manhattan.

New York City now covers approximately 200,000 acres of which eleven percent is landfill. True to their national genius, the original Dutch settlers began early to landfill swamps and marshes on the southern tip of Manhattan. By the late 1880s, however, the expansion of the city and of its commerce rapidly caused a shortage of office space in lower Manhattan. This raised the value of land to such an extent that only tall buildings could return a profitable income. The skyscraper type of architecture, which had just been developed in Chicago, thus immediately found extensive application in lower Manhattan.

In principle, the skyscraper architecture is based on the "steel skeleton" type of construction by which the entire weight of walls and floors is borne and transferred to the foundation by a framework of metallic columns and beams. It is probable that the hard bedrock, which underlies lower Manhattan, facilitated the erection of very tall buildings by this technique. In any case, the phenomenal rapidity with which lower Manhattan became crowded with skyscrapers is a symbol of its success as a commercial center and especially as a banking center.

The process began with the ten-story building at 50 Broadway, completed in 1889. One year later, the World or Pulitzer Building reached twenty-six stories on Park Row. In 1908 the Singer Building reached forty-seven stories and 186.5 meters in height at 149 Broadway. In 1913 the Woolworth Building at Broadway and Park Place not only became the world's tallest building with sixty stories and 237.4 meters in height but established a new style by incorporating gothic forms in its structure to give the impression, as mentioned earlier, that it was the equivalent of a cathedral of commerce. The Woolworth Building retained the record of height until 1929 when the Chrysler Building reached seventy-seven stories and 318.8 meters on 42nd Street at Lexington Avenue. But this record was soon bettered by several other skyscrapers, in particular by the 102-story, 380.4-meter Empire State Building at the corner of 34th Street and 5th Avenue in 1930. The twin towers of the World Trade Center, completed recently, now dominate the skyline of Lower Manhattan with their 107 stories.

I first saw Manhattan in October 1924 as I arrived from France on the steamer *Rochambeau.* Many trips to and back from Europe by sea until the late 1940s, and numerous crossings by ferryboat to and from Staten Island have given me the opportunity to recapture often and in its full force the impression of visual poetry that I experienced on first seeing the skyline of lower Manhattan.

Approaching Manhattan by ocean liner or ferryboat, the traveler first perceives the island as an ethereal body emerging from the water and ascending into the sky. The illusion creates different moods, depending upon the atmospheric conditions and the hour of the day. At times Manhattan seems to float in mist or in clouds as if it were a craggy mountain on an ancient Chinese scroll. Often it shimmers in precious pinks and delicate blues, one of William Blake's visions materialized on the shores of the New World. At dusk or at night, it glows like a luminescent body, a flaming torch. Seen from afar, Manhattan appears like a celestial city that never evokes in me material wealth or brute forces.

Yet the towers of Manhattan that so deceptively convey from a distance the ethereal splendor of a dream world are in fact a confusing and overpowering mass of steel and stone on hard bedrock. They are, furthermore, the creations of wealth and arrogance. They were erected not as a contribution to a unified shared endeavor, but as a display of individual power and pride. How could such uncoordinated ambitions engender so ethereal a silhouette against the sky? Through what mysterious alchemy did the random interactions of gross individual strivings bring into existence, in less than half a century, one of the most unexpected and grandiose architectural symphonies of the world? The quality of the Manhattan skyline is a perplexing expression of the fact that human beings can convert crude appetites into the splendors of civilization, and can transmute greed into spirituality. Repeatedly throughout historical times, they have created better than they had planned or even imagined. Works of universal and lasting value have often emerged from efforts aimed at the satisfaction of crude material urges.

Entering Manhattan by the Brooklyn Bridge also evokes an ambivalent response to the creations of humankind. Both sides of the bridge are crowded with tense human beings, often aggressively bent on exploiting each other for the sake of money, and obsessed with the pursuit of sensual pleasures. As judged from their behavior, many of them appear unaware of the sky and of the unique features of the natural and humanized environments linked by the bridge. They seem undisturbed by the brutality of the noise, by the harshness of the artificial lights, by the uncouthness and the anonymity of the human encounter. Automobiles and trucks generate a deafening vibration on the pavement of the bridge, creating a metallic hell in which human beings seem condemned to move mechanically and endlessly. Yet the bridge itself is like a gigantic diaphanous web flung across the sky, indeed like a lyre on the strings of which light plays at all hours of the day and of the night, singing the poesy of industrial civilization.

The massive pillars of the Brooklyn Bridge call to mind ancient temples in which profound mysteries were once enacted. This illusion is even more profound at night, when Manhattan sparkles and vibrates through millions of illuminated windows, each one of them a symbol of a passionate struggle for power and for creation. There also comes to mind the memory of J. A. Roebling, the architect of the bridge who died from a wound contracted during the first stages of the construction, and of his son Washington Roebling who continued the task. Washington Roebling's health broke down under the crushing load of his responsibilities, yet, though paralyzed, he continued to supervise the work from his bedroom window. The Roeblings' lives, dedicated to the construction of this poetical masterpiece of steel and stone, symbolize humankind at its highest—more concerned with worthy creations than with health and comfort.

Although Manhattan was the national capital, then the capital of the state of New York for only a very few years, it became the capital of the world during the twentieth century, first in the minds of countless human beings, then officially as the home of the United Nations. It achieved its prominence not only through wealth, political power and the diversity of its cultural endowments—including the diversity of its architectural styles—but by becoming the symbol of the hopes and excitement associated with modern life.

Each famous city has its own unique set of attributes that determine its public image. Manhattan has been a dreamland for countless millions of people all over the globe for more than a century. These people did not dream of it as a place for peace and comfort but as one in which to play their luck—an area of freedom and of unlimited possibilities where they had a chance to convert their ambitions into notable achievements. While skyscrapers are the results of selfish attempts to achieve power and material wealth, the city does not evoke only brute force, probably because its vertical, soaring lines express attempts to escape from crude matter into the sky.

Instead of being the outcome of conscious planning, the subtle complexity of the New York City skyline has emerged from the impact of an immense variety of people progressively adding to Manhattan the many natural settings of the five boroughs. The New York experience, although purely local, thus provides a lesson for the globe as a whole. While great diversity in any situation increases the complexity and numbers of problems, it also generates original and enriching solutions for the very problems it creates.

All great cities are made up of different human groups but usually become famous only after having homogenized their various ethnic subsocieties. This has not been the case for New York, not only because the ethnic structure of the city is extremely complex and still constantly changing, but because Manhattan achieved world leadership at a time when a very large percentage of its people had retained their original ethnic and cultural identities.

The person who invented the expression "melting pot" had never set foot in this country. If he had lived here, he would have discovered that what holds the people of New York together is not homogenization in a melting pot but, paradoxically, certain patterns of behavior that permit them to live *apart* from each other whenever they want—either as individuals or as groups. Rather than having lost their cultural identity in a melting pot, New Yorkers are instead like components of a human mosaic; they can be fitted together when the need arises, as they did during the 1965 blackout, but they usually prefer to function as independent units. They do not let the excitement of participation interfere with their desire for individuality and for privacy.

Human diversity has, of course, many drawbacks, but it also has beneficial consequences. It creates social tensions which lead to a search for attitudes and for laws designed to give equal rights to all citizens—irrespective of religion and of race, of age and of sex. Human diversity makes tolerance more than a virtue; it makes it a requirement for survival.

The coexistence of different cultural groups leads, furthermore, to the emergence of self-contained, competitive units that enhance the strength of the community as a whole. The WASP culture is the better for being complemented by the black culture, the Irish wit for being complemented by the Jewish wit, the Protestant religions for being complemented by the Roman Catholic worship of the Virgin and of the saints. Social evolution proceeds most rapidly when different cultures come into close contact with each other and thus can exchange information and goods, even though each retains its originality. The existence of numerous human groups within New York also greatly facilitates and enriches the contacts that are essential to economic and cultural growth.

People come to New York from all over the world to share in the exuberance of its urban life, but the natural environments of the city are just as diverse as are its humanized environments. New York has 930 kilometers of waterfronts along the ocean and along its riverways—the Hudson River, the Harlem River and the East River. No other city can match it for the diversity of its waterscapes. The diversity of its geological foundations is also phenomenal, as illustrated by the variety of woodlands and wetlands that exist in each of the five boroughs. According to the seasons, New York often provides the climate of the Equator or of the North Pole, with every degree of weather in between—not to mention the countless microenvironments created by the various types of buildings and by pollution.

Like human diversity, environmental diversity makes life in New York bewildering and traumatic, but it provides an enormous range of experiences from which all persons can choose so as to create and nurture their own self-selected personas. Environmental diversity does not make for comfort, but it helps human beings to discover who they are, what they can do and what they want to become. Individual New Yorkers can participate in an endless flow of public events in order to escape from themselves but they can also find shelter in anonymity if they want the solitude essential for creation.

The world is approaching a time when travel difficulties will make it imperative to find pleasures close to home. New York has been so generously endowed by nature with diverse landscapes and waterscapes that it could become a world leader in this trend. It could create within its own city limits an immense variety of quasi-natural environments suitable for almost every kind of activity and mood, thus providing stages on which to act out one's chosen life-style.

While I have emphasized the advantages of diversity, I am aware of the difficulties it creates. Diversity is at the origin of many conflicts, and it tends to make the world of things and the world of people inefficient and inconvenient. But I believe that, in the long run, diversity is preferable to efficiency and convenience, even preferable to the serenity of absolute peace. Without diversity, freedom is but an empty word; persons and societies cannot continue to evolve. Human beings are not really free and cannot be fully creative if they do not have many options from which to choose.

Human and environmental diversity results in painful problems of change and development; problems which are, in fact, occurring almost everywhere in the modern world, but usually later and with less intensity than in New York City. By trying to find a solution to the problems of diversity, New York acts as an experimental city for the rest of the world. Experimentation does not provide comfort, but rather the hope and excitement of discovery. This hope and this excitement have made New York a place that incarnates the challenges and the dreams of the human adventure.

Whether in the Netherlands or in Manhattan, or wherever human beings have profoundly transformed the surface of the earth, they have often created new, valuable cultural environments, but they have also destroyed or spoiled natural environmental values. In the course of their growth and evolution from one generation to the next they have developed technological and social structures that have become so large and so complex that the human mind cannot fully apprehend them or comprehend them, let alone manage them properly. I have mentioned earlier some of the dangers created in the Netherlands by excessive industrialization and urbanization. New York City presents even more unpleasant features associated with modern civilization, but I shall limit myself to a very few of which I have a direct experience.

I live in Manhattan in a huge apartment building. Looking at the city and its surroundings in all directions from my twenty-ninth-floor windows, I see everywhere the maddening automobile traffic and the dehumanizing banality of much recent skyscraper architecture. I often dream of what a wonderful human habitat Manhattan could have become if it had been developed more "humanly." In 1947, for example, the writer Paul Goodman and his brother, the architect Percival Goodman, published a book about urban planning, entitled *Communitas,* in which they illustrated by text and drawings how Manhattan could have been built in such a way that each street would open a vista onto one of the waterfronts and thus make walking a pleasurable experience almost everywhere in the city.

Environmentalists are justifiably concerned about nuclear warfare, radiations from nuclear plants, air and water pollution, shortages of natural resources, the destruction of the tropical rain forests, desertification and other forms of loss of agricultural land, but the greatest danger to our urban and technological civilization may well be its mechanical and social complexity. The confusion of traffic in and out of New York City, the bewildering flood of information and regulations pouring into and out of huge anonymous office buildings, the immense variety of unessential goods and artifacts which social conventions compel us to use, usually without rhyme or reason, are but symbols of ways of life that do not enrich human life and that no philosophy or computer can make intelligible to the human mind.

Complexity and enormous size commonly have paralyzing effects on many aspects of our activities including the creativeness of institutions. Most of the really new and original technologies of recent decades have come not from megacorporations but from small enterprises or small groups of investigators. In many cases, furthermore, great size and complexity make it more difficult for institutions to deal with fairly simple problems of daily existence that could be readily solved in a smaller context—witness the administrative difficulties presently experienced in the renovation of the Manhattan waterfronts.

Having had repeated contacts over many years with the official city institutions and with the citizen organizations involved in this particular problem of the waterfronts, I know that failure to act on them or the great delays in action are not caused by ignorance, indifference or negligence but by administrative and financial difficulties inherent in the complexity and huge size of the urban structures. By contrast, two important cases of environmental improvement of public waterfronts were rapidly set in motion by private initiatives.

Jamaica Bay, a large Atlantic bay adjacent to John F. Kennedy airport but within New York City, used to be so polluted as to be the most degraded area of the urban environment. Yet it has been restored to such an excellent state of ecological health that it is now the richest bird sanctuary on the Atlantic coast. One of the most interesting aspects of this renovation process is that it was begun by a minor civil servant, Mr. Herbert Johnson, the son of a gardener, who on his own initiative, without official instructions, made it a practice to plant suitable grasses, shrubs and trees on the islands of garbage in the bay. The growth of these plants attracted bird life and this eventually encouraged the urban authorities to develop more elaborate plans for the saving of the bay.

At the present time, one of the exciting environmental projects in New York City is the renovation of the Bronx River and of its shores. The Bronx River is in good shape as it runs through Westchester County and where it enters New York City through a gorge in a magnificent hemlock grove located in the grounds of the New York Botanical Garden. From then on, however, the river and its banks within the city present one of the worst examples of environmental degradation. Fortunately, a private organization called the Bronx River Planning and Action Group has undertaken to revitalize the river and its banks south of the Westchester County–New York City line. The program, if successful, will ultimately enrich the life of some 500,000 people who live in that part of the Bronx. Furthermore, it could create some of the most spectacular urban sights in areas that are now grossly degraded, especially at Hunts Point where the Bronx River empties itself into the East River. Most important probably is the fact that the renovation project is the product of local initiatives and has succeeded in involving the black and Puerto Rican communities who live in the neighborhood of the Bronx River. Financial help is now provided by the city and by private foundations but there is no doubt that individual initiatives were crucial for the initiation of the project and its early development.

THE GLOBE-TROTTER AT HOME

I have spent countless hours in international airports, in North and South America, Europe, Asia, Australia and even Africa. Most people think that international air travel will inevitably increase global homogenization and they are right of course, but only in a limited sense. New York, London, Paris, Frankfort, Stockholm, Rome, Athens, Sidney, Moscow, Tokyo are linked by the same kinds of aircraft operating according to the same international rules for taking off, navigating and landing. But even a blind person can perceive differences between the airports of the various cities, because their individual human atmospheres reflect the national conditioning of the preaviation era. These differences are even more striking when one compares international airports in the Pacific islands. Tahiti, Fiji and Hawaii have much in common with regard to climate, topography, resources and colonial history; furthermore these islands are now populated by a similar diversity of people—Polynesians, Malaysians, Orientals, Caucasians and other ethnic groups. But despite geographical and ethnic similarities, and despite the technological uniformity of international air travel, who can possibly fail to recognize the French influence on Tahiti, the English influence on Fiji, the American influence on Hawaii.

Hasty globe-trotters and scholarly travelers have good reasons to be impressed by the standardization of technologies and ways of life throughout our globe, but they are wrong in assuming that their own interests and experiences are those of greatest importance in the everyday life of local residents. Few are the people who want to live according to the habits and tastes of the international jet set. Ordinary people are usually eager to use some of the techniques and products of international technology, but in their own ways within the context of their local customs. Television sets are much the same all over the world but in any given year, the love songs they present and the facial expressions of the singers differ profoundly as one moves from Sidney to Athens and then to London or Paris; the differences are even greater between Jersey City and Guatemala City. The world is becoming a global village to the extent that several of its megasystems are linked by electronics, but the real life of people takes place in small neighborhoods where the news transmitted by global means of communication is far less important than that learned by local gossip at the barbershop or the hairdresser.

Scientists—whether specializing in the hard or the soft sciences—tend to be chiefly interested in worldwide problems, because they think that these provide information relevant to the discovery of general laws valid for all humankind. The trends toward uniformity are sufficiently real to give the illusion that the particularities of each place or social group are of less scientific interest than the generalities about these groups. In most cases, however, the manifestations of diversity are more important in everyday life than are generalizations derived from uniformity because they provide the materials out of which individual persons create what they prize most—the uniqueness of their own individuality and of their place.

Many people much of the time find it helpful to take advantage of technological standardization. The globe-trotter—whether motivated by restlessness, curiosity, or the search for knowledge—can travel over large parts of the world without changing habits. In most international centers he can start the day with ham and eggs for breakfast or with brioches and croissants if he is so inclined; he can dictate letters to a secretary in an air-conditioned office; he can get current quotations on the stock market and immediately wire instructions to his broker; he can receive information regarding current developments in politics or in technology; buy souvenirs made in Hong Kong or Taiwan; for dinner have roast beef with French wine or sushi with tea and sake; end the day with Scotch, brandy or vodka while discussing the American elections, any one of the liberation movements, or the new trends in writing, music and painting with people who have, like him, obtained sketchy information from one of the international magazines.

From almost any spot on earth, the globe-trotter can start on the next leg of his trip with the confidence that he will find, within 150 kilometers, jet planes staffed with polyglot professionals serving international dishes, drinks, cigarettes and souvenirs, all presented with international smiles. The only significant difference between airlines may be that the Japanese attendant will walk a few times through the aircraft in a colorful kimono, the French attendant will offer food and drinks with a Comédie Française accent, and the American attendant will wear a jazzy uniform the style of which is likely to change every year. Our globe is becoming, indeed, more and more standardized when we observe it as we fly from one continent to another for the sake of contacts with international groups.

The earth loses much of its global village mood, however, when one follows travelers on their way back home. In the John F. Kennedy airport, both the Japanese and the French businessmen or scholars wear the same kind of conventional business suits and carry interchangeable attaché cases. The Japanese businessman will keep his Western suit as he reports to his office in Tokyo but will change to a kimono and eat Japanese food with chopsticks when he reaches home. The French businessman probably had eggs with bacon for breakfast and drank cocktails before his meals while in the United States but will shift to petits pains, croissants and very black coffee in the morning, then have wine with his other meals whenever he stops in France. Global thinking and behavior are essential for travelers who move about the world to deal with global problems, but the more personal and pleasant aspects of individual lives are likely to be associated with daily activities peculiar to the place of permanent residence.

While human beings are essentially similar from the biological point of view, national and regional characteristics are the outcomes of historical accidents which have caused the human groups of a given area to be conditioned for long periods of time by a certain set of environmental and social forces. At the end of the world wars nationalism was discredited and almost rejected as an irrational and destructive force rendered obsolete by international technology. But while it is true that the abstract concept of nationalism is in retreat, nations survive and indeed continue to play an essential role, probably because they cater to a fundamental human need for kinship that used to be satisfied by the tribal system.

In his essay "The English people," George Orwell asks rhetorically, "Do such things as national cultures really exist?" To which he answers, as did Samuel Johnson when asked about the existence of free will, that this is one of the questions for which scientific knowledge is on one side and instinctive knowledge on the other. One does not need to adopt the Hitlerian doctrine that nations are based on racial differences and on a *Verbundenheit mit dem Boden*—a rootedness in the soil—to accept as an empirical fact the existence of different national, regional, and ethnic cultures. National feeling is first and foremost cultural, a state of mind based on a community of past experiences, of present interests and tastes, and generally of ill-defined but nevertheless influential aspirations. Americans are as hostile to the thought of eating horsemeat as are French people to the thought of eating cornbread. Chinese and Japanese painters and architects are proud of imitating the great masters and creations of their past whereas Caucasians prize originality in all forms of art. The English aspire to a form of democratic freedom which does not seem to have much appeal for the Russian masses.

In practice, most of the members of a given society want to act as they are expected to act. In part this is because it makes life easier, but in greater part because people have been conditioned to like what is considered desirable in their society and therefore find local types of behavior personally rewarding. Thus, while there is no reason to believe that all members of a nation share qualities derived from blood relationships or from natural characteristics of the land on which they live, plain observation reveals that a distinctive array of intellectual and behavioral attitudes is associated with the adjectives Mediterranean and Nordic, and even more with the adjectives American, English, French, German, Greek, Italian, Russian and Spanish. Likewise the words Chinese and Japanese denote tastes and behaviorial attitudes that have remained different for centuries regardless of political regime. The American social critic Max Lerner was not entirely facetious when he wrote that in England everything that is not forbidden is permitted; in Germany everything is forbidden unless it is permitted; in France everything is permitted even if it is forbidden; and in Russia everything is forbidden even if it is permitted. Granted that these statements are oversimplifications, they are nonetheless true enough to illustrate that groups of people whom the accidents of history have forced to live together for several generations tend to share a body of ideas, values and tastes which govern their lives.

National characteristics rarely develop as a result of pressure from the outside. They evolve spontaneously as a structure of relationships generated by the constant interplay of the forces that are at work in a given country. They are the expressions, not of race or climate, but of human choices based on the collective acceptance, either willful or through social pressure, of certain conventions and myths. In his essay, "The English People," George Orwell stated that "myths which are believed in tend to become true because they set up a type of 'persona' which the average person will do his best to resemble." According to Orwell the courageous behavior of the British population during the Second World War "was partly due to the existence of the national 'persona'—that is to their preconceived idea of themselves." Nations need heroes as concrete symbols of what they imagine themselves to be.

Despite its claims to internationalism, the Bolshevik regime was intensely national from its very beginning. As early as the mid-1920s, it had developed a deep concern for the early masters of Russian literature and followed Lenin's precepts that the cultural values of the past would provide the foundations for the cultural values of the future. This national attitude had deep roots in Russian literature. Turgenev for example used to speak of the "deep unfathomable chasm" between the Russian conception of social problems and the views held on the same subject by French, English, German, and other Europeans. Even the anarchist aristocrat Prince Peter Kropotkin, who had long affirmed that differences regarding social justice and order existed only among the middle classes of the various nations, came to realize that the workers also saw things in quite different ways depending upon their nationality.

The concern for the quality of Russianness which inspires almost all Russian writers has been vividly expressed by the Nobel Prize winner Solzhenitsyn even though he has been discredited by the Kremlin for his political views. In the first chapter of his novel *August 1914,* Solzhenitsyn presents Russian nationalism not as an idea but as an affective reaction, deeply rooted in the subconscious. Russianness has something to do with the soil, and everything to do with the people, their beliefs, and their language. It is unrelated to political ideologies and is as natural as a tree, deeply rooted in the subsoil of a culture and rising toward the heavens in a desire to transcend the immediate situation. For Solzhenitsyn, *Russia is not a place on the map but an image shaped by the Russians* [italics mine]—much as England was for George Orwell the cultural environment for an attitude formulated by the English people themselves.

In fact, a territory once occupied by a given people may be lost without this causing the breakdown of national identity. Such persistence of identity is well recognized in the case of the Jews; but it has also been observed in many other people; for example, the Yaquis, the Navajos and the Cherokees have survived as a people even though they have been displaced from all or part of their tribal territories. National individualism also commonly survives political domination. The people of the Republic of Ireland feel and express a continuity with the Irishmen of more than a thousand years ago. They were part of Great Britain for a long period of time but steadfastly rejected any thought of identification with Englishmen. There is evidence furthermore that tragic experiences tend to reinforce the identity of social groups or nations, and indeed help people to maintain their conception of themselves and of their collective identity in a wide range of physical and sociocultural environments. Except in a limited biological sense, it is not the actual past that shapes the view we have of ourselves and that generates the rules of our behavior; it is rather the image we create of the past.

Aleksandr Blok, who was one of the best known Russian symbolists of the early twentieth century, tried in his essay "The Collapse of Humanism" to differentiate between civilization and culture. For him, civilization emphasizes material possessions, calendar time, and the growth of specialties. Culture, in contrast, is the "musical" reality uniting spirit and flesh, humankind and nature; it is a truly elemental force. "Great is our elemental memory . . . the musical sounds of our cruel nature have rung in the ears of Gogol, Tolstoy, Dostoevsky." Vague as it is, this distinction between civilization and culture helps to elucidate why international technology has not yet destroyed national individualism. It explains also some of the forces that have made the national spirit a powerful creative force.

All over the globe, human beings have now shaped the land on which they live. They have replaced most of the wilderness with farmlands, pastures, gardens, parks, which have become so familiar that they are commonly assumed to be of natural origin. Wherever this influence has been intelligently applied, humankind and nature have entered into a symbiotic relationship which modifies both of them, creating thereby the characteristics of each region and each nation and thus giving its form to each particular civilization. The earth, or any particular place on it, can be viewed only as a physical system for the support of life, whereas the words "place" or "nation" denote an environment which has been emotionally transformed by feelings.

There are, of course, many expressions of nature that have not thus been humanized. When the flow of natural events is undisturbed it generates types of wilderness which commonly far exceed humanized landscapes in emotional power. In our daily life, however, we rarely function in the wilderness and are hardly ever passive in front of the natural world. We fence it, manipulate it and later use it to create environments suited to our needs but also and even more to our traditions and aspirations. By inserting our dreams and our sense of order into ecological determinism, we shape the raw stuff of nature into patterns which integrate the materials provided by the wilderness with those of our human nature—a truly creative symbiotic process.

Since most present landscapes reflect certain kinds of human intervention, they could be otherwise than they are but this does not mean that they could be almost anything. To be viable, humanized environments must be compatible with ecological constraints. It is certain, however, that a given environment in an established society is usually the product of a long process of adaptation, with the ultimate result that the society makes the environment a dimension of itself. If people live in a place long enough, the quality of the place enters into the substance of their lives.

The relationship between landscape and humankind can be considered a true *symbiosis* because it involves biological forces that bring about creative changes in both components of the system. But there is also a large element of conscious choice in most interventions of humankind into natural systems. Social groups, like individual persons, never *react* passively to environmental situations; instead they *respond* to them in a purposive manner. It has long been recognized, as mentioned earlier, that the growth of civilization is favored by variable and challenging environments—whether the challenge comes from topographic, climatic, or social stimuli. Civilizations, however, do not emerge from the passive reactions of the social group to these stimuli; they are purposeful responses made in an attempt to create some chosen ways of life.

I have repeatedly mentioned in earlier parts of this book how profoundly and irreversibly I have been conditioned by my early experiences in France. I might have enlarged on the French tendency to emphasize the local aspects of life by quoting Voltaire when he made fun of Professor Pangloss's generalizations about the world's problems and expressed through Candide's voice his own opinion that we must first take care of our own garden. But I shall instead present images and statements from another French writer, Antoine de Saint-Exupéry, a pioneer in French aviation whose life and writings illustrate how one can be concerned with universals yet give to this concern a characteristically French expression.

Shortly before his death on a flying mission at the end of the Second World War, Saint-Exupéry wrote to one of his acquaintances in France that he had always wanted to travel, but never wanted to emigrate. "J'ai appris tant de choses chez moi qui ailleurs seront inutiles" (I have learned at home so much that would be useless elsewhere). And he went on to write, "Si je diffère de toi, loin de te léser, je t'augmente. Tu m'interroges comme on interroge le voyageur" (If I differ from you, far from doing damage to you, I enrich you. You question me as one questions the traveler). This was one of the many statements by which Saint-Exupéry conveyed his conviction that the best way for each person and each place to contribute to the world is to affirm its own identity.

Saint-Exupéry's messages in his book *Le Petit Prince* are of such universal significance that the book has been translated into many languages and is still widely read today. Yet what could be more French than the affirmation by Saint-Exupéry that a particular rose, a particular fox and a particular landscape have values that do not come from their intrinsic qualities but from the fact that the little prince had taken care of them and had made them part of his own being.

Americans extol the virtues of the wilderness. In contrast, many French people tend to find a richer meaning in those aspects of nature that they have humanized or more exactly, to use Saint-Exupéry's word, that they have "apprivoisées." Dictionaries translate the verb "apprivoiser" as "to tame" but there is much more to "apprivoiser" than simple taming. In *Le Petit Prince,* Saint-Exupéry has the fox tell the boy, "On ne connaît que les choses que l'on apprivoise, dit le renard.

Les hommes . . . achêtent des choses toutes faites chez les marchands. Mais comme il n'existe pas de marchands d'amis, les hommes n'ont pas d'amis. Si tu veux un ami, apprivoise-moi." In the published English translation of *The Little Prince,* this passage reads as follows: "One only understands the things that one tames, said the fox. Men . . . buy things already made at the shops. But there is no shop where one can buy friendship and so men have no friends any more. If you want a friend, tame me." As used by Saint-Exupéry "apprivoiser" does not mean simply taming. It implies shared emotional experience, mutual understanding and appreciation. In fact, the fox gives the little prince detailed instructions concerning the rites through which their acquaintance can become a true friendship—a kind of relationship that is intensely "local" since it involves only two participants who develop patterns of behavior to express their unique relationship.

Each country and ethnic group has its own traditional system of intimate relationships and of rites not only among people but also between people and the land. It is this traditional system that makes globe-trotters feel and behave differently when they return home and makes them exclaim as they reach their destination, "This is the place. There is no place like home," because the most unique and important aspects of one's individuality are precisely those that have been shaped by the environments in which one has developed and functioned.

4

TREND IS NOT DESTINY

THE BEAUVAIS SYNDROME

THE MATERIAL POWER OF SPIRITUAL FORCES

WORLD TRENDS AND CONTEMPORARY GLOOM

SOCIAL ADAPTATIONS TO THE FUTURE

4

TREND IS NOT DESTINY

THE BEAUVAIS SYNDROME

The histories of Manhattan and the Netherlands illustrate how various human groups have evolved socially and technologically along different channels to solve problems originating either from their activities and aspirations or from natural conditions. Manhattan and the Netherlands also illustrate that civilizations usually develop to the point of the absurd some of the attitudes, practices and techniques that were responsible for their initial success. Skyscraper architecture is a convenient way of dealing with space shortage and provides visual excitement in modern cities, but it often causes physiological trauma and esthetic boredom. The use of the motorcar was at first a pleasurable way to increase freedom of movement but has become a dangerous social addiction. Fortunately trends that appear suicidal can be interrupted before irreversible damage has been done. While all civilizations appear to be mortal, some of them, Phoenix-like, are capable of being reborn from their ashes.

Gothic architecture provides a well-documented example of a breathtaking technique that was once carried too far for safety. The first great achievements of Christian architecture were in the Romanesque style. Wonderful churches and monasteries of that early period are still surviving today and it is probable that Romanesque architecture could have evolved further along its own lines of development if it had not been for the influence of an extraordinary man known to history as the Abbé Suger.

Suger was born in 1080 or 1081 near Saint Denis, five kilometers from my native village, Saint Brice. He was educated in the church, joined the Benedictine order of monks, and, at an early age became a trusted counselor of the French kings, first Louis VI, then Louis VII.

He was elected abbot of the great Benedictine Abbaye of Saint Denis in 1122, and acted as regent of France while King Louis VII was away on the Second Crusade. Suger was of small stature and is reported to have been a physical weakling, but was a man of immense energy with an administrative genius that he applied both to his religious responsibilities and to the government of France.

The Abbaye of Saint Denis had been founded in the seventh century, then rebuilt and rededicated by Charlemagne in 775. Around 1135, Suger undertook to replace it with a much larger building of a different architectural style. This new Abbaye, which still exists today, marks the first transition on a grand scale from the Romanesque to the Gothic style. It immediately became a religious landmark and was used as the burial place of the kings and of their families until the French Revolution.

Some 32 kilometers north of Saint Denis and some 16 kilometers east of the village of Hénonville where I was raised, the small city of Senlis began in 1155 the construction of a cathedral which was for a while, along with the Abbaye of Saint Denis, the finest achievement of early Gothic architecture. From then on, a multiplicity of cathedrals and smaller churches in the Gothic style emerged all over Europe, especially in France. While all these religious edifices derived their architectural inspiration from Saint Denis and Senlis, the various cities tried to outdo each other in the size and height of the cathedrals and in the boldness of their architecture. The vault of the nave in Notre Dame of Paris, completed in 1163, had the record height of 30.5 meters. This record was broken by Chartres in 1194 with 34.75 meters, then by Rheims in 1212 with 38.1 meters, then by Amiens in 1221 with 42.7 meters. Competition between cities was at least as strong a motivating force in architectural design as was the glorification of God.

The nave of Amiens cathedral was so high that it gave a sense of insecurity, but its splendor and boldness created a spirit of jealousy among the citizens of Beauvais, a small town south of Amiens located 16 kilometers north of the village of Hénonville. In 1247, Beauvais began building its own cathedral with the intention to raise its vault 4 meters higher than that of Amiens. Furthermore, Eudes de Montreuil, architect of the new cathedral, dared to increase the feeling of openness and the penetration of light by decreasing the number and the thickness of the pillars and buttresses. Light could thus flood the interior from windows that were remarkable for their narrowness and 18.3-meter height. The choir was completed in twenty-five years but the vault crashed in 1284, twelve years after its completion. During the next forty years, the choir was strengthened, but at the cost of some of its ethereal quality. Construction was interrupted during the Hundred Years War with England but was taken up again in 1500. The gigantic transepts were begun and a lantern tower was raised in 1552 over the transept cross to a height of 152 meters. The Beauvais cathedral was then the largest, highest and boldest structure in the world, but it collapsed in 1573 on Ascension Day (Holy Thursday), fortunately a few minutes after the crowd of faithful had left the cathedral for a procession through the city. The medieval architects had developed unjustified confidence in their skills and had pushed Gothic architecture beyond the point of safety. The disaster spelled the end of this type of architecture and the Renaissance style replaced it in all types of buildings, including cathedrals and smaller churches.

Many other civilizations have similarly developed excessive confidence in the social and technological proficiencies which had accounted for their initial success. In ancient China, the disciplined and sophisticated scholarship of officials long contributed to the quality of government, but eventually led to mandarin paralysis. In France, the centralization of authority under the kings of divine right made for great power in both national and international politics, but eventually led to the Revolution. In our times, the obscene consumption of energy and resources, the senseless multiplication of motorcars, the ever-increasing complexity of urban agglomerations are the equivalents of the Beauvais syndrome. They will inevitably lead to disasters if Western civilization continues to choose material growth as its dominant value. History proves that technological proficiency, economic prosperity, and political organization have never been sufficient to assure that a society will remain successful and even persist. Human institutions must be held together by cohesive forces of a spiritual nature. The power of such spiritual forces accounts for the survival of many ethnic groups despite centuries of political subjugation—as is the case for the Jews, the Irish, the Amerindians, the Basques, etc. Spiritual forces also are a dominant factor in the rise and fall of social and religious institutions, indeed of nation-states and empires.

THE MATERIAL POWER OF SPIRITUAL FORCES

At the time of Jesus's birth, Rome was the administrative and financial center of an immense empire. Augustus had been given consular status, the title of princeps, and the power of the tribune. He constantly traveled over the empire to gain direct familiarity with its problems. He restored social order in the land, reorganized the army, strengthened the frontiers, reformed taxation and the administration of the law, patronized the arts and literature, and helped to define Roman architectural forms. Livy, Ovid, Horace and Virgil flourished under his rule. At least as important as his reforms in the material domain, however, was the new prestige that he gave to the old pagan religions associated with patriotic values. This policy decreased the intensity of conflicts in Italy and increased the spiritual significance of life. Testimony to the soundness of the administrative and social structures created or encouraged by Augustus is the fact that they endured and often functioned well for three centuries after his death.

During Augustus's time and largely thanks to him, the empire was approaching its most prosperous and most stable phase. Writing of the Antonine age, Edward Gibbon stated that the human race had never been happier than during that period, approximately a hundred years after Augustus's death. But there was another side to this picture—the evils of slavery, the pauperization of urban people, the concentration of privileges in the upper classes—all symbols of a deficiency in spiritual values which probably had a large part in the eventual acceptance of Christianity by people from many different social classes—including those who benefitted from the prevailing state of affairs.

The civil servants of the Roman Empire were highly skilled in engineering and organization; they brought food, water, raw materials, and manufactured goods from all parts of the then known world. They created immense buildings, roads and aqueducts, many of which still exist today and some of which are still in use. In fact, they developed a technical civilization which went almost as far as could be done before the invention of the steam engine. But this was almost exclusively a materialistic civilization which did not satisfy spiritual longings.

As the empire grew, so did the threats to it, both from the interior and the exterior. There was disaffection within Italy from the less favored provinces and social classes. The heavy dependance on imports of food and raw materials created financial problems, especially as Rome's granaries in North Africa became less productive due to mismanagement of the land. Outside Italy, the barbarians from northern and eastern Europe were never absolutely controlled. Rome's problems had become so formidable at the time of Marcus Aurelius's death in 180 A.D. that in the words of the senator and historian Dio Cassius, the age of gold had turned into the age of iron. Probably most important was the progressive loss of the spiritual values that had first made for the greatness of Rome. The empire was destroyed from the outside by the barbarians but chiefly because it had been weakened inside by the loss of the sense of pride and duty that had enabled Rome to master the Mediterranean world and part of western Europe.

Jesus was born approximately eight years before Augustus's death, at a time when the empire seemed indestructible. At that time also the synagogue ruled supreme over Jewish life in Palestine with regard to both religious beliefs and social behavior; the values enshrined in it appeared as permanent as those enshrined in the Roman law. Jesus's followers had no social organization except for the practice of their faith and a consensus of values, but their spirituality gave them a strength that enabled them to survive hostility and persecutions, to enlarge their communities and eventually to replace the existing social order. Who could have imagined during the period of the Pax Romana that the Roman Empire would collapse under the blows of the barbarians, and that the barbarians themselves would rapidly accept the teachings of the Cross?

Nothing is known of Jesus's life except for the few facts recorded in the New Testament. Most important, probably, was his decision to sacrifice himself for the sake of spiritual values. He could have elected not to come back to Jerusalem and therefore avoid the trial that led to his crucifixion but he exposed himself deliberately, thereby symbolizing that commitment to a cause is often more powerful than obvious determinism in shaping the course of events.

The story of Islam also shows that spiritual values can be more powerful than material forces in shaping history. Mohammed was born in Mecca around 570 A.D. and spent his early days in poverty as a shepherd. Although he eventually became a tradesman he formed the habit early in life to withdraw periodically into the mountains to meditate and pray. In the year 610, at age forty, he had a vision during which the angel Gabriel revealed to him the words of God and instructed him to memorize them and teach them to other human beings. Mohammed did not begin preaching until 613 when, in typical Beduin style, he reported his spiritual experience by recitatives, packed with vivid comparisons and exhortations. He proclaimed the existence and all-powerfulness of the one and only God, creator of the universe, a God of mercy as well as of justice.

He made a few converts in Mecca, the first being his own wife, but was rejected by the local community. He therefore moved to the oasis of Medina where he made converts who enabled him to gain control of Mecca in January 630. By the time of his death in 632, practically the whole Arabian peninsula had been converted to his doctrine of monotheism and to the rules of conduct that had been dictated to him during his visions. Mohammed's biographers agree that his influence was not due to doctrinaire teaching but was exerted through his moral ascendency and political shrewdness. His teachings were in the form of recitations—*qur'am* in arabic from which we have derived the word Koran. In his recitations, he often referred to "islam" as meaning "surrender to God's will" and this became the name of the religious faith he preached and established.

Mohammed's followers continued to extend the faith of Islam by arms with such vigor that, within a century after his death, they had created an empire that extended from the Pyrenees in France to the Pamirs in Central Asia. Spain, North Africa, Egypt, the Byzantine Empire south of the Taurus mountains and the Persian Empire were thus welded together into a political unit that stretched 4800 kilometers from east to west, embraced a great diversity of peoples and regions and rivaled the size of the Roman Empire at its peak.

The Arabs managed to keep control over the lands they had conquered for a century but the different parts of the empire tried to regain political autonomy as soon as the Arabs began to lose the passionate spirituality that had been instilled into them by Mohammed. Although most (but not all) of them retained the Muslim faith the empire itself was transformed into a world of distinct and often warring states, conscious of a common identity based on religious faith but divided by tribal or racial loyalties. When the Mongols overran the Muslim world in the Middle Ages, the original Arab empire had long ceased to exist because it had lost the spiritual passion originally given it by Mohammed.

The phenomenal ability of the human spirit to overcome material forces and to shape events is most readily recognized in historical situations identified with a particular person, as is the case with Jesus, Mohammed or Gandhi—a person who confronts a problem then decides to act even if it means self-sacrifice. In many cases, however, social and political innovations result not from the overpowering influence of a single person but from changes in limited groups of people who create a new mood and thus prepare the ground for action by the masses.

The modern world, for example, has been largely shaped by the philosophers of the Enlightenment who formulated the rationalistic view of nature and of humankind that eventually led to the development of technological civilization and to the acceptance of constitutional government. In our century, the emergence of Israel as an independent state and of Japan as a westernized nation were also the consequences, not of spectacular individual feats, but of collective changes of mood.

Many heroes played a role in the emergence of the state of Israel but the Zionist movement itself has a diffuse genesis. Zionism originated in eastern and central Europe during the late nineteenth century, from the cultural attachment of the Jews to Palestine where one of the hills of ancient Jerusalem was called Zion. This ancient feeling for Palestine was intensified when anti-Semitic policies convinced certain Jewish intellectuals that their ideals could be realized only in Palestine, their historic homeland.

The first agricultural settlements of Palestine were founded in 1882 by a small trickle of Jewish youth; additional ones were financed by the Baron Edmond de Rothschild of Paris; there were twenty-two settlements in 1900, forty-seven in 1918. From then on, the Zionist movement proceeded, but along several different social philosophies. Certain Zionists wanted to take advantage of the agricultural settlements to create a really new society based on socialistic or even communistic principles; many early kibbutzim came close to fulfilling this ideal. Other Zionists did not believe in the desirability of creating an independent Jewish state but wanted to establish a Jewish cultural center, almost anywhere in the world, which would serve as a place for the regeneration of Judaism and for the spread of its spiritual influences. Still other Zionists emphasized the religious aspects of the enterprise and insisted on the strict observance of religious laws in Jewish life. Finally, there were and still are many Jews who are hostile to the very principle of Zionism and to the establishment of a separate Jewish state.

The anti-Semitic atrocities of World War II probably made inevitable the emergence of the new state of Israel. Whatever its ultimate fate, its phenomenal success in many sociocultural and technological aspects of life demonstrates once more that human determination and will can overcome almost all natural obstacles. Israel is the seat of a multiplicity of simultaneous social experiments ranging from the communistic kibbutzim to the capitalistic industries of Haifa; or from the conservative rigidity of Hasidism to the adventurous spirit of the great universities. The simultaneous occurrence of all these social experiments in a small area with a small population and limited resources gives assurance that humankind has not lost its ability to choose, imagine, act and thereby create.

The modernization of Japan began in the nineteenth century with changes that returned political power, then largely held by feudal lords, to the imperial throne. This revolution took place during the reign of Mutsuhito who took the name Meiji ("enlightened rule") in 1868. The Meiji leaders realized that Japan did not have the military power to prevent Americans and Europeans from penetrating their country, trading with its people and thus changing Japanese ways. Recognizing that Western strength depended on constitutionalism, national unity, industrialization and military force they sent missions to study various aspects of governmental and technological life in America and Europe. Political reforms came rapidly and the government initiated a program of industrialization that was soon transferred to private investors.

Efforts were also made to replace feudal with national loyalties. To this end, the Shinto cult was given a high position in the political hierarchy in order to supplement Buddhism with the cult of national deities. The system of universal education established in 1873 was used to develop a body of "ethics" largely based on Shinto ideology and Japanese ways of life. Loyalty to the emperor hedged about with Confucian teaching and Shinto reverence thus became the center of social ethics.

The success of the Meiji revolution is evident in the industrial growth of Japan and the even more spectacular recovery from the disasters of the Second World War. But it is not clear why Japan has been so much more successful than other Asian countries in assimilating and developing the scientific and technological aspects of Western civilization. A tentative explanation is presented in a book that was immensely popular in Japan a few years ago, and that has been translated into English under the title *The Japanese and the Jew.* According to its author Isaiah Ben-Dasan (said to be a Jew who was raised in Japan), some peculiarities of the Japanese climate have required the development of strict practices of cooperation among Japanese farmers for the successful cultivation of rice. These practices introduced into Japanese life a cohesion and discipline that facilitated a rapid shift of the work force to technological activities requiring a high level of organization and standardization.

Other aspects of the success story are presented in a more recent book, *Shinohata: A Portrait of a Japanese Village,* written by R. P. Dore, an Englishman who has long lived in Japan. Until 1960, Japanese village life was spartan, with very little mechanical equipment. Then in a few years, farm work was motorized and village life filled with electric appliances. According to Dore, the foundations for this revolution were laid in the 1890s and 1900s when the Meiji rulers decided to use the savings of Japanese economy, not for agriculture or the welfare of the people, but for industrial development. Technological prosperity was therefore achieved at the cost of the farmers, who did not begin to derive benefit from it until the 1960s and 1970s. Such a policy, which ignored for so long the needs of the plain people, would probably not have been possible under usual democratic conditions. It required the highly centralized structure of the Meiji political system and the discipline of Japanese life. The economic and technologic consequences of the Meiji revolution are symbolized in Japan by the geometrical aspect of much modern architecture and by the rigid structure of professional life which often appears as crude regimentation to the foreigner.

Japan is almost as much a human creation as are the Netherlands. Nature has not been generous to the country and history has not always been kind. But the Japanese have brought good out of bad, carving rice paddies from steep hillsides, reclaiming land from a difficult sea, building strong walls and levees to hold back flooding rivers. They began improving their small, rocky islands centuries ago, and each generation has passed to the next the responsibility of accomplishing more. They have built the fastest and best railroad system in the world and they are now in the process of completing a network of railroad tracks that will unite all the islands of Japan, at even greater speed than that of the famous bullet train. As in Holland, however, this technological way of life may be approaching its limits. Many a Tokyo commuter who expends almost a fifth of his day and even more of his energy in a madcap shuttle reaches home daydreaming of his boyhood in a country town.

National discipline and organization are not of course the only distinctive features of Japanese life and culture. There is the seeming madness of Pachinko players, behaving as if they were mesmerized gamblers, but there is also the reverence of the crowds walking through the temples, shrines and parks and the sensuous richness of the Shinto celebrations. Despite modernization, there still is the exquisite austerity of life in many homes.

When my wife and I left Japan after an extensive visit a few years ago, the many people who had looked after us gave us as a last present a toy representing the small caldron used to cook over the fire in old villages. This was to thank me for having emphasized in my lectures that spiritual human values can give charm to life without the benefit of advanced technologies.

Several aspects of modern Japan justify the hope that, when a human society elects to adopt technological civilization, it can also continue to cultivate attributes which come from innate spiritual values rather than from the rational brain, and which have their origin in the depth of the ages.

WORLD TRENDS AND CONTEMPORARY GLOOM

The most distressing aspect of the modern world is not the gravity of its problems; there have been worse problems in the past. It is the dampening of the human spirit which causes many contemporary people, especially in the countries of Western civilization, to lose their pride in being human and to doubt that we shall be able to cope with our problems and those of the future. At a meeting on "Ethics in an Age of Pervasive Technology," held in Israel a few years ago, the American philosopher Max Black went so far as stating, "I happen to think that the problems raised by technological advance are probably insolvable [sic]." Such a feeling of hopelessness has historical precedents. In *Five Stages of Greek Religions* (1953) Gilbert Murray traced the fall of the Greco-Roman civilization to a "failure of nerve." History shows, however, that other societies have experienced dark days in the past, yet have managed to recover—as illustrated in our times by Germany and Japan.

I am as much disturbed as anyone by the thousand devils of the present social, technological and environmental crises. In fact, I am inclined to believe that we shall remain on the brink of catastrophe for at least two or three decades, if only because of the likelihood of acute, even if temporary, shortages of energy and other resources. I also realize that several aspects of the present world problems make them quantitatively and qualitatively different from those of the past. For example:

A. Today's problems are no longer isolated and confined to small population groups.

B. Many deleterious agents are spread over most of the globe, as in the case of radioactivity and of acid rains.

C. Useful technological innovations tend to have unexpected consequences as when the widespread use of pesticides to control insects brings about dangerous changes in the food chains of birds, fishes and eventually humans.

D. There is an unprecedented interconnectedness of effects; the political conflicts in the Middle East influence petroleum production and thereby American living patterns as well as the attempts of poor nations to develop more productive agricultures and industry.

There are great tragedies in the world today. Paradoxically, however, much of contemporary gloom originates not from the difficulties we are actually experiencing, but from disasters that have not yet happened, and may never happen. We are profoundly disturbed by the possibility of nuclear warfare and of really serious accidents in nuclear reactors, but disturbed also by the unproven hypothesis that the widespread use of fluorocarbons from spray cans will damage the ozone layer and thereby expose us to dangerous levels of ultraviolet radiation. We are collectively worried because we anticipate that world conditions will deteriorate if population and technology continue growing at the present rate. The earth will soon be overpopulated and its resources depleted; there will be catastrophic food shortages; pollutions will rot our lungs, dim our vision, poison us, alter the climate and spoil the environment. The spread of income between the have and have-not nations will widen even further. This will certainly increase terrorism and may eventually lead to the use of nuclear weapons, if only as a form of blackmail.

During the past few decades most writings by sociologists, economists and environmentalists have expressed a pessimistic mood about the future. The volumes on *Limits to Growth* published by the Club of Rome in 1972 deserve special mention in this regard because they were the first to provide a seemingly scientific basis for the atmosphere of gloom that now prevails over much of the world. They have been read, or at least quoted as gospel truth, by millions of people who have accepted the doomsday forecasts of mass starvation, depletion of resources, overwhelming pollution and political chaos sometime during the next century. Many similar publications have recently appeared. All of them take the form of computer models of what the future will be, either in the world as a whole, or in parts of it, constructed on the basis of existing data concerning population, resources, pollution and the present trends in these and other demographic, social, economic and technological categories.

The last and most monumental contribution to this kind of world game is an eight hundred-page *Global 2000 Report to the President* prepared in the United States by the Council on Environmental Quality and the State Department in collaboration with thirteen Federal agencies. The purpose of the *Global 2000 Report* is to determine the "probable changes in the world's population, natural resources, and environment through the end of the century" to serve as foundation for long-range planning. At the very beginning of the report its authors acknowledge the difficulty of obtaining reliable information for such an ambitious endeavor. In their own words, "The executive agencies of the U.S. government are not now capable of presenting the President with internally consistent projections of world trends in population resources and environment for the next two decades." A few examples will suffice to illustrate the unreliability of the information which serves as a basis for quantitative studies about the present state and the future of the world.

According to official statistics issued by the English Ministry of Agriculture, Fisheries and Food, the total amount of foodstuffs recorded for 1976 in Great Britain were so low as to make it appear that British people eat on the average much less than the minimum recommended by the U.N. Food and Agricultural Organization. Yet, practically all British people were well nourished at that time and not a few of them were in fact overweight. In Great Britain, as in other parts of the world, people use in their diets and in most of their activities many items not entered in official documents. Since the statistical records in Great Britain are vastly better than those in most if not all other parts of the world, the *Global 2000 Report* and similar global studies probably justify the phrase "Garbage In, Garbage Out" that has been coined to criticize the use of unreliable information for the design of world models in many social studies.

The lack of convincing information concerning the earth reserves of fossil fuels such as petroleum, natural gas, coal, peat is obvious from even the most casual reading of newspapers or magazines, and uncertainty is even greater concerning the prospects for so-called renewable forms of fuel, directly or indirectly derived from the sun.

One can assume with the authors of the report that the consumption of nonfuel minerals will continue to increase and that shortages of some of the rarer ones may occur, but it was shown during a recent international convention of geologists that new techniques—from remote sensing to geochemical methods—are revealing the existence of vast reserves of essential minerals in many parts of the American continent.

Deforestation has certainly reached tragic levels, especially in tropical and semitropical areas, but the statement in the Global Report that "if present trends continue the forest cover . . . in the less developed regions (Latin America, Africa, Asia and Oceania) will decline by 40 percent by 2000" needs to be supplemented by the information that large programs of reforestation are going on in many parts of the world, for example in North Africa, the Sahel, Ethiopia, India and, on a particularly grand scale in the People's Republic of China.

For centuries, large areas of plains and hills in China have been essentially treeless as a result of a process of deforestation that began thousands of years ago. In 1949, however, the People's Republic began a program of reforestation which, according to official statistics, has now extended to some seventy million acres and continues at the rate of some ten million acres per year. These figures seem compatible with the accounts of Westerners who have recently flown over China. A special aspect of this reforestation program is the creation of a "Great Green Wall" that will extend from the river Amor in Manchuria to the high semidesertic plateaus of Cansu province over a length of almost 3200 kilometers and that will eventually cover more than 200 million acres. This project has already begun in areas of some twenty million acres especially threatened by erosion. Whereas the historical Great Wall of China was built in the third century B.C. to protect the Celestial Empire against the Mongols, the new "Great Green Wall" is being established to protect the People's Republic against ecological disasters.

The futility of trying to make useful projections about the future state of the world appears in an almost comical form in the section of the *Global 2000 Report* that deals with problems of health. The Report states that the average life expectancy at birth for the world as a whole will *increase* by eleven percent from 58.8 years in 1975 to 66.5 years in 2000 "as a result of improved health," but it states also that since a longer life expectancy will cause crowding and decrease living standards, the ultimate result of improvements in health will be an increase in the incidence of disease and of mortality in many parts of the world. Anyone familiar with the uncertainties and complexities of public health trends will take these statements with a big grain of salt.

There is nothing new in the general conclusion of the Report that "if present trends continue, the world in 2000 will be more crowded, more polluted, less stable ecologically and more vulnerable to disruption than the world we live in now. . . . Despite greater material output, the world's poor will be poorer in many ways than they are today." Such grim warnings have been repeated ad nauseam since the doomsday forecasts of the Club of Rome's *Limits to Growth.* Like many others, I also believe that our present form of technological civilization will eventually collapse *if* present trends continue . . . but what a big *if* this is.

Using techniques and types of information similar to those used to prepare the Club of Rome's *Limits to Growth* and the U.S. *Global 2000 Report,* Latin American scientists arrived at conclusions radically different from those reached in these two studies by asking different questions. Instead of projecting future conditions on the basis of present policies and trends, the authors of *The Latin American World Model* asked "How can global resources best be used to meet basic human needs for all people?" Because *The Latin American World Model* does not start with the concept "If present trends continue" but assumes instead that certain social and technological changes can, should and almost certainly will take place in several places, it projects a future in which basic human needs could be satisfied in Latin America and Africa within a relatively short time. *The Bariloche Foundation World Model* which was worked out as a counterpart to the Club of Rome *Limits to Growth* and was published in Canada under the title *Catastrophe or New Society* emphasizes that *Limits to Growth* was modeled to fit the conditions of the northern industrialized countries. If the same questions had been asked for the underdeveloped southern countries, the answers would have been very different because the limits to growth in these regions are not physical but sociopolitical and for example depend largely on policies of land ownership.

The authors of the *Global 2000 Report* find in their computer models evidence for "a vicious circle . . . that leads to starvation and economic collapse by midcentury" for Asia as a whole. This might be true if present trends continue but the agricultural, technological and economic "miracles" that have happened during the past few decades in Japan, South Korea, Taiwan, the People's Republic of China make it clear that trends have often been reversed. It is probable in fact that, in the modern world, rare are the trends that continue very long.

Human beings are not likely to remain passive witnesses to situations that they regard as dangerous or unpleasant. Their interventions may often be unwise, but they always alter the course of events and make mockery of attempts to predict the future from extrapolation of existing trends. In human affairs, the consequences of determinism are always less probable, less interesting and usually less important than the actual events because these are largely brought about by deliberate human choices and activities. In my opinion, industrial societies have a good chance of surviving and even of remaining prosperous because they are learning to adapt to the future.

SOCIAL ADAPTATIONS TO THE FUTURE

We adapt to heat, cold, crowding, poverty and other environmental and social conditions when we experience these conditions and minimize their effects by appropriate changes in our physiological mechanisms and our ways of life. The phrase "social adaptations *to the future*" therefore sounds nonsensical since societies have not experienced the conditions, largely unpredictable, to which they will have to adapt in years to come. Our personal biological adaptations, however, are more subtle than would appear from the first statement of this section. In ordinary life, our minds and our bodies make adaptive responses to situations that have not yet occurred, but that we anticipate. For example, our heart starts beating faster at the mere thought of having to run in order to catch a train at some time in the near future; our secretion of various hormones is increased when we know that we shall soon have to face a special situation, even one as mild as delivering a lecture to an unknown audience. In a similar way, human societies can adapt to the future, even a distant one, by anticipating the probable effects of situations they are likely to encounter in times to come, and by taking in advance adequate measures in the light of these anticipations.

Until our times, most important changes took the world by surprise. Consequently there was no possibility of affecting their occurrence and it was difficult to control their manifestations. Now, in contrast, the possible effects of technological and social innovations are discussed long before they become manifest, especially if they are likely to be dangerous. We try to imagine the "future shocks" that humankind will experience when its ways of life and its environments are altered, at some undetermined time in the future. However, the very fact that certain symptoms of "future shock" have been described in advance leads us to make anticipatory mental and social adjustments to them, with the result that the symptoms are not likely to occur, at least not in their predicted form. Steps are now being taken, for example, to protect us from the evil doings of Big Brother and from other social calamities predicted by George Orwell for 1984. Anticipation has of course always influenced human activities, but it is only during recent decades that a significant number of important anticipations can be based on a wide range of reliable information and at times on precise scientific knowledge.

Anticipating the likely consequences of natural processes and of human activities is quite different from predicting the future. The future cannot be predicted for two different reasons. One is that prediction would require complete knowledge of the past as well as of the present which is impossible. The other is that human beings practically always impose a pattern of their own choice on the natural course of events. Admittedly, there are aspects of the future that are largely predictable because they follow logically and inevitably from antecedent conditions and events. For example, one can predict with a high degree of accuracy the amount of water that will accumulate in a reservoir built in a geological formation of known properties, behind a dam of known dimensions thrown across a stream of known average flow; one can also almost predict how long the reservoir will remain useful as it fills up with silt and other products of erosion. In contrast to the possibility of predicting such forms of logical future one cannot predict whether the dam will or will not be built, and if it is, exactly where and when. Such decisions involve innumerable human choices—of technological, economic, esthetic and even ethical nature. In 1913, President Theodore Roosevelt and a significant percentage of the Sierra Club membership approved the building of a dam across the spectacular Hetch Hetchy River in Yosemite National Park, to create a water supply and a source of hydroelectric power for San Francisco. I doubt that this project would have been approved by President Jimmy Carter and by the present membership of the Sierra Club. More recently, the American people have been willing to create the Hoover dam on the Colorado River but have rejected any thought of interfering with the Colorado as it flows into the Grand Canyon.

Thus, the "logical" future imposes constraints on human activities but there is a "willed" future that is first imagined and decided in human minds and that comes into being only through systematic planning and efforts. The optimists, among whom I try to be, are those who believe that the willed future based on humanistic values can be successfully integrated with the effects of natural forces and with the social structures emerging from scientific technology. The following examples illustrate that the willed future is bringing about, in many situations, desirable changes which are based on anticipations of effects and events that have not yet occurred. In other words, social adaptations to the future are taking place now.

The North American continent has never had a high population density but it might have become overpopulated if the high birth rates of the past had continued for a few more decades. The writings of demographers, and in particular Paul Ehrlich's immensely popular book, *The Population Bomb,* created a widespread awareness of the threats to the quality of life posed by uncontrolled population growth. Thus, while there is still much empty space and unused resources in the United States and Canada the anticipation that North America could become overpopulated in the *next century* has significantly contributed to the decrease in average family size, even in Harlem and in Catholic Quebec. Birthrates are now so low in many social groups that the North American continent may achieve Zero Population Growth some time next century, and this is likely to occur perhaps even earlier in several European and Asian countries.

Environmental degradation became a lively public issue only two decades ago, and most antipollution programs are less than ten years old. Yet, certain aspects of environmental quality have already been vastly improved as a result of control measures initiated during the 1960s and 1970s, especially in Europe and North America. The levels of social pollutants have decreased in many large American and European cities—indeed even in Tokyo. Several streams and lakes that were so grossly polluted as to be qualified "dead" during the 1960s have been brought back to a level of purity compatible with a rich and desirable aquatic life. Since I have described in detail and analyzed many of these environmental achievements in my book *The Wooing of Earth* I shall limit myself here to mentioning one example from Europe and one from the United States.

Ever since the beginning of the Industrial Revolution and until the late 1950s, London was the most polluted large city of the Western world. As a result of the control measures taken by the London City

Council under the Clean Air and Clean Water Act of 1957, the annual amount of sunshine over London has now increased by some fifty percent, there has not been a single case of "pea soup" smog during the past ten years, the songbirds mentioned in Shakespeare can once more be heard in the city parks, and salmon—that most fastidious of fish—has returned to the Thames. In New York City, Jamaica Bay used to be grossly polluted with garbage and sewage but thanks to a variety of antipollution measures it is now in such good condition that fin fish and shellfish, including oysters, are sufficiently abundant to support a fishing industry. Furthermore the bay has become a rich bird sanctuary and a most attractive part of the Gateway National Recreation Area, recently established around the New York–New Jersey coastline.

Forests that had been devastated are being allowed to recover and massive programs of reforestation are being carried out in several parts of Africa and Asia, and especially in China. Areas that had become desertic are being protected against goats and other animals and can thus reacquire a diversified flora and fauna—for example in Israel and even in some parts of North Africa and the Sahel. While environmental degradation is still increasing in many parts of the earth, particularly in the form of desertification and of destruction of the tropical rainforest, there are signs that several modern societies are learning to work with nature.

An interesting aspect of environmental improvement is that, in all cases, measures against environmental degradation were taken long before the situation had become desperate. The air of cities had not become poisonous to a significantly dangerous level before legislation was formulated and applied to control urban air pollution. Lakes and rivers were far from being "dead" at the time when control of water pollution was begun. It was anticipated, however, that air and water pollution would reach unacceptable levels if the emission of pollutants was allowed to continue at the past rates for one or two more decades. The measures to improve the environment were therefore taken not as a response to *actual* emergencies, but rather to the *anticipation* of emergencies.

Adaptation to the future is also apparent in the case of supplies of raw materials. It was recognized a few decades ago that shortages of some metals were in sight. Long before there was any real shortage, however, scientists established that, with appropriate technological changes the functions served by the metals in short supply could be fulfilled by other more abundant metals or by some synthetic products. Admittedly, the use of substitutes commonly entails greater expenditure of energy but adaptation to the future is taking place even in energy production.

The prices of all forms of energy have greatly increased during the past decade but there has not yet been any real shortage except as a result of social or political disturbances such as labor conflicts or the 1973 Arab oil embargo, or when snow makes it impossible to move fuels where they are needed. There are still enormous reserves of various fossil fuels in several parts of the world but they may not be available because of technical or sociopolitical difficulties.

As soon as it was recognized that the reserves of petroleum are not unlimited and will not last more than a few decades, industrial countries began to build nuclear reactors as sources of energy. However, geologists soon pointed out that the natural reserves of uranium also are limited and would soon be exhausted if nuclear energy were being produced on a large scale by conventional nuclear fission. Nuclear technologists therefore focused their efforts on the development of the breeder reactor which produces nuclear fuel while generating energy. Research in this field was given high priority in the U.S. and Europe. During the 1960s and 1970s, however, several groups of scientists and citizens took a stand against the breeder program, not necessarily because they doubted its feasibility, but because they were alarmed by the dangers inherent in its operation—in particular the inevitable accumulation of plutonium.

Concern for the potential dangers of nuclear fission technologies has begun to affect official policies. In Europe as well as in the U.S., an increasing percentage of research funds are being devoted to energy sources that used to be neglected—solar radiation, wind, waves, tides, biomass, geothermal, and the so-called synfuels which are various forms of oil and gas derived from coal gasification, coal liquefaction, shale oil retorting and tar sands conversion. It is worth noting that, while the production of synfuels is still in the experimental stage, more is known concerning their potential effects on human health and on the environment than was known concerning the effects of coal and petroleum until two decades ago. The recent evolution of policies for research in energy production has thus been a continuous adaptive process determined not by present shortages or the occurrence of accidents but by the anticipation of possible shortages and of accidents that have not yet materialized.

In addition to its preventive aspects against anticipated dangers, adaptation to the future has begun to take a more *creative* character. For example, the phrase "good environment" is no longer taken to mean only freedom from noxious influences; it implies also the creation of surroundings having esthetic and emotional values. City planners used to be concerned almost exclusively with public health facilities, transportation systems, and efficiency in the various aspects of urban economic problems. They are now beginning to emphasize, in addition, the role of neighborhood parks, of pedestrian streets and of "city centers" in the furtherance of social and cultural activities. The fear that urban agglomerations will soon become unmanageable because of their excessive complexity is also encouraging attempts to divide them into smaller communities easier to apprehend and to manage. Creative adaptations to the future are the chief mechanisms by which human beings can invent new social structures on a rational basis.

The ability to anticipate long-range consequences does not mean that modern societies will necessarily be able or willing to act early and vigorously enough to prevent deleterious effects. Pessimists have good reasons to believe that someday, somewhere, a social or technological innovation will be carried so far so fast that it will cause irreversible damage to the human species or to global ecology. However, while a catastrophe following overshoot cannot be ruled out, there is reason for hope in the wonderful resiliency of natural systems and of human beings. Advances in knowledge will facilitate anticipating the long-range consequences of technological and social innovations and may thus enable us to overcome the myth of inevitability.

A biological problem which has recently caused a great deal of public alarm will serve as a last example, and perhaps as a caricature, of the social process denoted by the phrase "adapting to the future." The new scientific techniques of so-called genetic engineering make it possible to modify the hereditary constitution of microbes. From this scientific fact, many persons have concluded that it will eventually be possible to modify also the genetic constitution of all other living organisms, including the human species.

The genetic engineering of human beings is not scientifically possible now, and many are the biologists who share my view that it will never be possible to a significant degree because the complexity of higher organisms implies a high level of integration which would be profoundly disturbed by an important change in any one of their constituents. Yet several institutes of bioethics have been created in which physicians, biologists, sociologists, jurists, theologians are assembled to discuss the medical, ethical, legal and theological aspects of modification in human nature that might eventually be brought about by methods of genetic engineering—yet to be imagined! Thus, we try to adapt not only to the future which is being created by scientific technology, but also to the future imagined by science fiction!

5

MATERIAL RESOURCES AND THE RESOURCEFULNESS OF LIFE

FROM THE WILDERNESS TO HUMANIZED NATURE

RAW MATERIALS AND RESOURCES

THE MERITS OF ENERGY SHORTAGES

CREATIVE ADAPTATIONS AND ASSOCIATIONS

5

MATERIAL RESOURCES AND THE RESOURCEFULNESS OF LIFE

FROM THE WILDERNESS TO HUMANIZED NATURE

There are two very different kinds of satisfying landscapes. On the one hand are the various forms of wilderness which have evolved without human interference and in which all forms of life have spontaneously achieved ecological equilibrium among themselves and with the physicochemical environment. On the other hand, there are landscapes and waterscapes that have been transformed by human interventions to satisfy human needs and tastes in harmony with natural forces. I shall not discuss the wilderness, in part because I have had little personal experience of it and also because it has been described, analyzed and illustrated in many excellent publications. I shall emphasize instead relationships between human beings and natural forces because most of what we call nature is really the outcome of this interplay which expresses itself in the innumerable forms of humanized environments.

The first and last chapters of *Silent Spring,* the famous book by which Rachel Carson made the general public aware of the dangers of pesticides, begins with pictures of imaginary towns, the streets of which lead toward a rolling landscape of meadows, grainfields and hills crowned with woodlands. These pictures were derived from Rachel Carson's memories of her youth in western Pennsylvania early in this century. In her book, she describes an enchanting scenery in which every form of life was "in harmony with its surroundings. . . . The town lay in the midst of a checkerboard of prosperous farms, with fields of grain and hillsides of orchards. . . . The countryside was famous for the abundance of its birdlife . . . ; the streams flowed clear and cold out of the hills and contained shady pools where trout lay. *So it had been from the days many years ago when the first settlers raised their houses, sank their wells, and built their barns* [italics mine]."

Rachel Carson's make-believe countryside has widespread if not universal appeal. Most of us have an esthetic ideal of nature which is probably the consequence of the visual conditioning created in us during the Stone Age by life in a savanna environment. Fortunately, the combination of woodland, open space, waterscape and horizon, which is characteristic of the savanna, is also compatible with many different types of agricultural sceneries. It can be achieved in the classical French landscape style, in the more romantic English treatment of the land, in the complex and symbolic design of Oriental parks—and also in the tremendous natural sceneries of the American national parks.

We are only now beginning to acquire accurate information concerning what people really find attractive in scenery. Tests carried out with different social groups have revealed an almost universal preference for orderly landscapes, in which "nature" has been tamed and even disciplined. Most of us long, not for the wilderness itself, but rather for a taste of *wildness* in the humanized sceneries we admire.

The rural areas Rachel Carson had known in her youth were not "natural" environments; they had been created out of the primeval wilderness two or three centuries ago. The settlers of Pennsylvania had not only, in her words, "raised their houses, s[u]nk their wells and built their barns." In order to create open land for their farms and towns they had been compelled to cut down much of the forest that initially covered western Pennsylvania. Most of the American countryside, like that of Europe, has been created by farmers, just as gardens are created by gardeners.

In fact, human beings have created artificial environments out of the wilderness over most of the earth, wherever they have made their homes. In Europe and Asia, several of these "humanized" environments have remained fertile and esthetically attractive for centuries or even millennia and are now true homes for humankind. They have become so familiar to us that we tend to forget their origin; we contemplate them in a mood of casual acceptance and reverie without giving thought to the huge areas of primeval nature that had to be profoundly transformed, or destroyed, before they could fit our biological needs and also our esthetic longings.

Until the present generation, most human beings took great pride in their ability to transform the surface of the earth. At the end of the eighteenth century, for example, the English naturalist William Marshall stated that "Nature knows nothing of what we call landscape because this refers to habitats manipulated by human beings for their own purposes." Of Great Britain he wrote, "No spot on this island can be said to be in a state of Nature. There is not a tree, perhaps a bush, now standing on the face of the country which owes its identical state to nature alone. Wherever cultivation has set its foot, Nature has become extinct. . . . Those who wish for Nature in a state of *total neglect* must take up their residence in the woods of America [italics mine]." The expression "total neglect" seems to imply that Nature can achieve perfection and fully develop its potential values only as a result of human management. And it should be noted in passing that large areas of the American forest had already been converted into farmland at the time Marshall wrote.

Whereas people used to be proud of the changes human beings have introduced into nature, it is now fashionable, in contrast, to affirm that "Nature knows best" and that any human intervention into the environment is likely to have deleterious or even disastrous consequences. We have so completely talked ourselves into the belief that good environmental management is based on a few elementary scientific principles of ecology that we commonly use the phrase "ecological movement" to denote the modern environmental movement, as if the adjective "ecological" assured us of an environment satisfactory for human life and for the earth. The truth is, however, that most aspects of human life imply some conflict with natural ecosystems, and therefore involve disagreement with the teachings of textbook ecology.

Consider, for example, the pride that home owners take in the lawns they maintain in front of their houses, whether bungalows or mansions. The lawn appears to them as a tranquil and beautiful piece of "nature," but is in fact a monstrosity when considered from the point of view of theoretical ecology. Almost everywhere and especially in the temperate zone, the creation and maintenance of a lawn involves the expenditure of much energy and other resources to destroy and control the weeds, brush and trees that would naturally grow on it if not constantly eliminated, and that indeed reestablish themselves as soon as the lawn is neglected.

A flower or vegetable garden is also a most unnatural environment from the purely ecological point of view. Few of the plants cultivated in it could long survive in the wilderness. After being planted, garden plants can prosper only if constantly protected against weeds, insects, animals of every size, and countless other forms of life, as well as against natural forces that would destroy them or at least interfere with their growth.

Paradoxical as it may sound, the farmer's activities also imply the destruction of natural ecosystems. All over the world, farmland had to be created out of the wilderness by a variety of techniques and at great cost of energy—to cut down the forest, drain wetlands, irrigate deserts, and also to destroy many forms of wild animal and plant life as well as their native habitats. The success of crops and pastures and the maintenance of farmland quality require an endless fight against the native fauna and flora that would soon return in their wild forms if they were not controlled—often by violent means. Successful farming, like gardening, is incompatible with the ecological equilibrium that would exist under natural conditions.

Experience shows that alterations of natural ecosystems need not be destructive and may indeed be highly creative. The guiding principle for success is that artificial ecosystems should be as compatible as possible with the prevailing ecological characteristics of the regions in which they are created. Lawns, for example, have long been popular and highly successful in the British Isles because the local climate and especially the rainfall pattern are favorable to the establishment of turf. In contrast, good lawns are more expensive to establish and maintain in most parts of North America as well as in many other parts of the world.

The constant struggle against nature required for the maintenance of a lawn is the more difficult and costly in resources and energy because of the common desire to establish essentially the same kind of lawn everywhere—preferably with the characteristics of a championship golf course. Having observed for several decades the sufferings of the lawns on the grounds of Rockefeller University and of Central Park in mid-Manhattan, I wish botanists would cooperate with landscape architects to develop lawn grasses and other kinds of ground cover suited to different types of climate, rainfall, soil composition and human use.

In most parts of the world where farmers have used agricultural policies ecologically suited to natural local conditions, agricultural lands that were initially derived from the wilderness have commonly remained productive and have even become more fertile in many places. Certain contemporary practices, however, stand in sharp contrast to empirical ecological wisdom. In several parts of Texas and of the American West, for example, very high yields of various crops have been achieved by irrigating semidesertic lands with underground water (consisting in many places of fossil water) which must be pumped at great cost. Since the reserves of underground water are being depleted, and since irrigation water contaminates the soil with various salts as it evaporates, this type of farming will eventually have to be abandoned, leaving behind a legacy of degraded land. Fortunately, human beings can learn from experience as illustrated by the discussions going on at the present time concerning the future of farming in western Kansas.

Until a few decades ago, farmers believed that western Kansas was too dry and too hot to grow anything but wheat and feed grains that require little water. The situation changed, however, when it was discovered that corn could be readily grown in this semiarid region and give high economic returns by heavy irrigation with water pumped from the Ogallala aquifer of fossil water which underlies parts of the high plains states from Texas to Wyoming. The national significance of this kind of irrigated corn agriculture is indicated by the fact that forty percent of grain-fed beef used in the United States comes from cattle fattened in areas irrigated by water pumped from the Ogallala aquifer. The result of this massive use of irrigation, however, is that the water table has declined rapidly over the whole region, to such an extent that the reserves of underground water will be depleted in a very few decades.

In western Kansas, pumping from the Ogallala aquifer began in earnest only after World War II with the availability of high-capacity pumps and cheap natural gas to provide power. Use of the fossil water was akin to rich strikes of gold, silver or oil but, since the bonanza is being mined out, western Kansas will have to adapt to new conditions. In certain parts of Texas, the land has already gone back to sagebrush because irrigation has had to be discontinued. Similarly most of western Kansas would revert to buffalo grass and shrub, if it were not possible to plan for a progressive transition from irrigated corn agriculture to the raising of less water-intensive crops and, perhaps ultimately, to dry-land farming of wheat and grain sorghums.

The new practices of farming on a very large scale, with huge and complex mechanized equipment, have profoundly decreased the bucolic appeal of some rural areas. In the public mind, however, the degradation of the natural environment is chiefly identified with damages caused by the Industrial Revolution. As industrialization and urbanization began on a large scale in England, so did environmental degradation and the protests against the rape of nature. William Blake symbolized the belching factories by the expression "satanic mills." Ever since the eighteenth century, and especially during our times, more and more people have come to believe that industrialization is the one villain responsible for the destruction of environmental quality. But in reality, other changes in our ways of life have also played a large part in despoiling the charm of the countryside.

The small village where I was born, Saint Brice-sous-Forêt, lies on the edge of the forest of Montmorency, a few miles north of Paris. When the American novelist Edith Wharton settled in Saint Brice, where she lived from 1918 until her death in 1937, the plain between Paris and the village was occupied by highly productive truck gardens and especially by orchards of cherry, pear, and apple trees. The orchards were so numerous along the road that, in Edith Wharton's words, "we seemed to pass through a rosy snow storm whenever we travelled in the spring." I have traveled the road from Paris to Saint Brice on many occasions during the past thirty years, but with little pleasure because the gardens and orchards have now been replaced by a grey suburban area. Small lawns and a few flowering trees near the houses are lamentable substitutes for the wide-open views that could be had along the road when Edith Wharton lived there and for the "rosy snow storm" she enjoyed in the spring.

Suburban sprawl is not the only factor that has decreased the quality of landscapes around large cities. On a weekend in May 1945, my wife and I got off the train in Peekskill forty miles north of New York City and walked north on the Old Albany Post Road. This used to be the road linking Manhattan to Albany in the past, but most of it has now been replaced by Highway 9. We were walking on the few miles of the Post Road that still exist in their original form as a dirt road. The stretch we followed first ran through woodland, then reached a small intimate valley crossed by a lively little stream. There were gardens and meadows, apple orchards in bloom on the western slope, a few cows and, at each end of the valley, one house with farm buildings. For the urban dweller, this was "nature" in its most appealing mood. We bought some seventy acres of abandoned farmland, all of it extremely rocky except for the narrow alluvial area along the stream, and we built a fieldstone house on one of the slopes.

Today the scenery has greatly changed. The farmers who occupied the two old houses have died and in fact had discontinued farming several years before their death. The valley which was once intensely cultivated and supported many people now has only a few residents, who do not farm. Brush, trees and poison ivy have invaded orchards and meadows on the slopes and in the lowlands. Vistas remain open only where we, and two neighbors, try to control the wild new vegetation at great physical and financial cost. Just as the growth of anonymous suburbia brought about the disappearance of pleasant farming sceneries along the Paris–Saint Brice road, so the uncontrolled growth of brush and trees has decreased the bucolic charm of our Old Albany Post Road valley.

Writing of the farming scenery and great parks of England, Nan Fairbrother stated a few years ago, "The spacious and orderly planning, the serene vistas into the open country, these proclaim the assurance of a society confident in its power to dominate the surrounding world by its own intellectual force. . . ." In contrast, the recent changes along the Paris–Saint Brice road and the Old Albany Post Road symbolize the extent to which modern societies fail to maintain the visual and other sensory qualities of humanized environments. These are but two minor examples of the many types of scenic degradation which are brought about, not by William Blake's satanic mills or by modern industrial developments, but by changes in the ways of life and by the abandonment of agriculture.

Except for the real wilderness, the most appealing manifestations of nature are now found in farming areas and on large estates. But farming can rarely survive near large cities under present economic conditions, and most large private estates are also destined to disappear because of heavy taxation and also because they do not fit the tastes of wealthy people of the new generations. It has therefore become urgent to formulate land policies that might permit the maintenance of environmental quality around large urban areas; laws to this end are being developed in certain European countries. In parts of the Netherlands, for example, farmers are being subsidized for the maintenance of canals, hedgerows and windmills on their land; while such conservation practices may decrease somewhat the economic productivity of agriculture, they permit the survival of the cherished eighteenth-century landscape. In a similar spirit, Bavarian shepherds are being subsidized to graze their sheep in high alpine meadows; the long distances involved increase the cost of grazing, but the practice has ecological advantages because, when the grasses are not kept short, they are prone to be uprooted by thawing spring snows and to tumble down the mountains.

I have discussed in *The Wooing of Earth* many cases of artificial land management that have yielded desirable results. A comparison of the outcomes of two of the largest volcanic eruptions on record—in Krakatoa and in Santorini (also called Thera)—will illustrate still further how human intervention can bring about environmental values that would not have spontaneously emerged from natural processes.

In 1883, a stupendous volcanic eruption destroyed two-thirds of Krakatoa Island, situated in the Malay Archipelago between Sumatra and Java. The explosion had a violence comparable to that of a million hydrogen bombs and released enormous amounts of volcanic dust which spread throughout the whole earth's atmosphere. All living things were destroyed on Krakatoa, and more than 37,000 persons are known to have been killed on Krakatoa itself and in the neighboring islands, chiefly by the tidal wave that followed the eruption. When the eruption ended, what remained of the island was covered with a thick layer of lifeless lava. Progressively, however, wind and waves brought a variety of microorganisms, plants and animals to the shores and life took hold on the lava. Granted that the new ecosystem is inferior to that which existed initially, Krakatoa is once more covered with abundant animal and plant life derived from the Malay Archipelago, less than a hundred years after the eruption. Human beings, however, did not move back to the island except for short visits to observe periodically the progress of the returning living forms.

In the Aegean Sea, the island of Santorini was split 3500 years ago by a volcanic eruption estimated to have been four times more violent than that which occurred on Krakatoa. The remnants of Santorini were buried under a deep layer of pumice, in places 61 meters deep, and all life on it must have been destroyed. Human beings, however, did eventually return to the island; they have established agriculture and continuously lived on it ever since. Santorini has been devastated by other volcanic eruptions during historical times; as late as July 9, 1956 half of its 5000 houses were destroyed and fifty-three persons were killed. Despite these repeated natural disasters, Santorini provides what is perhaps the most spectacular illustration of how the human presence can enrich the creativeness of nature.

The ship on which I was traveling through the Aegean Sea a few years ago had reached Santorini late at night and my first view of the island was in the early morning. Even though I had read much about it, I was overwhelmed when I saw the perpendicular cliff, 305 meters high, plunging with its incredibly intense whiteness into the dark blue water 365 meters deep—testimony to the eruption which had split Santorini 3500 years before.

There is no water on the island except for that brought by the rains and stored in large underground cisterns; the soil of volcanic pumice is so poor that only a few shrubs and trees can take root in it and become established. Yet, cultivated fields quilt every patch of level land, and Santorini produces the best tomatoes and the most heady wine of Greece. As visible proof of cultural prosperity, white houses and colorful church domes gleam in the sun against the luminous sky on the very upper edge of the cliff.

The vastly different types of recovery following the volcanic cataclysms in Krakatoa and in Santorini bring into relief the contrast between the spontaneous blind processes of nature and the changes guided or imposed by human choices and decisions. After the 1883 explosion, life on Krakatoa became reestablished, but the recovery has been a matter of chance and has resulted in a poor imitation of the original flora and fauna. Left undisturbed, nature would similarly have reintroduced some of the common Aegean forms of life on Santorini, but whereas Krakatoa remained uninhabited, human beings returned to Santorini and resumed on it the process of social and cultural evolution. The Santorini inhabitants have experienced many ordeals but have succeeded in generating from the pumice and from the tormented Aegean rock new types of products and new forms of civilization. The small tasty tomatoes, the wine and the spectacular architecture of gleaming houses and colorful church domes on the edge of the cliff could not possibly have emerged without human imagination and perseverance.

RAW MATERIALS AND RESOURCES

In its original form, the Malthusian thesis was based on the theory that malnutrition and disease would inevitably result from continued population growth. Its more recent forms, as exposed for example in the books *Limits to Growth* and *Global 2000 Report,* emphasize in addition that the economies of industrial countries will be soon confronted with severe constraints resulting from shortages of many types of natural resources. This gloomy view of the future depends upon a multiplicity of assumptions concerning the supplies of certain raw materials on the surface of the earth, and the use of the resources derived from these raw materials.

The earth has always been poor in "natural resources." Substances become resources only after they have been extracted from the raw materials which contain them and have been manipulated to some human end. Farmland rarely exists as a natural resource; it has to be created by human labor from some form of wilderness. Aluminum does not exist as a free metal in nature; it must be separated by sophisticated chemical techniques from bauxite or some other aluminous ore. Petroleum and uranium are not useful as such; they become energy resources only after having been incorporated into complex technological systems. And so it goes for practically all so-called natural resources.

The phrase "reserves of resources" also leads to much confusion. For some experts the word "reserves" applies to the *proven* supplies of a given material whereas for other experts it applies to *absolute* amounts of that material on earth. The phrase "proven reserves" has economic connotations; it implies supplies known to be available within the time frame during which industries can operate under economically profitable conditions. In contrast, the phrase "absolute reserves" is a geologic concept, namely the total amount of a particular material in existence somewhere in the globe. For economic reasons, the efforts to discover absolute reserves are usually limited to the needs of the present time and of the immediate future. Estimates of absolute reserves are therefore extremely problematic. So far new reserves have usually been discovered whenever the need has arisen.

An extremely optimistic view of the resource problem appears in the anonymous statement made a few years ago in the course of a general discussion. "Since the entire planet is composed of minerals . . . the literal notion of running out of mineral supplies is ridiculous." Even granting that most of the earth's mass is beyond human reach, enormous areas of its crust and of the oceans are fairly accessible. Reports from the U.S. Geological Survey and the Battelle Memorial Institute assert that some four hundred trillion metric tons of usable materials are within the one-kilometer thickness of the earth's surface that can be reached by modern technology. Since the entire world consumption of sand, gravel, oil and metals is only of the order of twenty-two billion tons a year, "We should be able to support ourselves for at least 20,000 years—enough time, it would seem, to adjust our institution and population growth to the resources that are available." The cornucopians, in other words, equate practically the whole surface of the earth with "available" resources. More recently, this optimistic attitude has been reinforced by the hypothesis that many raw materials could be obtained from the moon and other planets, and could even be processed and tooled in outer space.

In practice, humans have used at any given time only those parts of the earth's surface which they could easily reach and manipulate. Stone Age people searched for rocks of the right kind that could be shaped into weapons and tools. They also used clay, fibers, wood and leather, all materials available almost everywhere. Metallic gold is readily separated from its ores and indeed often exists as pure metal in nature, but is too rare to lend itself to large-scale industrial applications. Copper was the first metal to come into widespread use because it can be easily reduced from its ores to the metallic form at low temperature. The usefulness of copper increased immensely after it could be alloyed with tin to produce bronze—the metal alloy of the first great civilizations. The Roman legions mastered the Western world with bronze weapons made with copper obtained from Spain and with tin from Cornwall.

Iron is much more abundant than copper but can be reduced to the metallic form only at high temperatures by complex technologies. Metallic iron, therefore, became available on a large scale only several centuries after copper and bronze. From then on, iron has been the chief metal for the manufacture of tools and weapons. Iron tools made it possible to transform most of Europe from an area of thick forests into fertile croplands in a few centuries during the Middle Ages.

Aluminum is even more abundant than iron but its preparation in the metallic state and its tooling present difficult technical problems and require large amounts of energy. Aluminum was not obtained as a pure metal until early in the nineteenth century and did not become an important technological resource until several decades later.

Iron accounts now for more than eighty percent of the total metal consumption in the U.S., aluminum for some ten percent and copper comes in third place. In the case of these metals, as well as of others, the first raw materials used for their production were naturally the ones most readily available and most easily processed. Prehistoric people first prepared copper from crystals of malachite that they could easily recognize and simply pick off the surface of the earth. When this source became scarce, ores of progressively lower copper content came to be used. Only ores containing at least five percent copper were considered worthwhile at the turn of the century, but contents as low as 0.4 percent or even lower are now economically acceptable. A similar downward trend in the acceptable richness of ores has taken place with regard to other metals, for example when industry shifted from the very rich magna type iron ore of the Mesabi range in Minnesota to the much poorer taconite. Aluminum is now produced from bauxite, but when this raw material becomes scarce there probably will be a shift to clay or other aluminous materials that are practically inexhaustible.

In theory, it is possible to continue using less and less desirable but more abundant ores. In the words of Harrison Brown, "Man could, if need be, live comfortably off ordinary rocks. A ton of granite contains easily extractible uranium and thorium equivalent to about 15 tons of coal plus all the elements necessary to perpetuate a highly technological civilization. Indeed, it would appear that we are heading for a new stone age." The rub, of course, is that the problems of environmental degradation, of energy consumption, and of capital costs become more and more severe as less desirable ores are used. While the top mile of the earth's crust contains an enormous amount and variety of useful mineral elements, it contains much more valueless or even dangerous materials which must be disposed of at great cost of capital and energy.

One answer to the resource problem is to consider the functions to be served and find proper substitutes. In current industrial practices few are the functions that cannot be performed by more than one material. Admittedly, the first substitutes selected may not perform the function as well as the metal they replace, but advances in technology continuously enlarge the range of interchangeable products. At the present time, for example, most electrical wire is made of copper. Aluminum could take its place for many present uses, granted that the change would require adjustments that might be somewhat disturbing. Optical fibers may come to be widely used in the telephone system and may even present advantages over copper because they can transmit many more bits of information in a much smaller volume.

Shortages of mercury could probably be managed because substitutes are available for each of the major uses of this metal. Furthermore, the toxicity of its compounds has led public health authorities to ban some of its former uses. Of the 2000 tons of mercury imported by the U.S. in 1968, the largest amount was for the production of caustic and chlorine, substances that can be produced almost as well by the diaphragm cell, which was used before the mercury cell was introduced, and which requires only common materials.

Among the most versatile and important substitutes are the various plastic materials synthesized not only from petroleum or natural gas but also from plant products. Plastics have in fact been found superior to metals for a variety of purposes and, as is well known, are increasingly finding their way into many aspects of domestic life and of technology. An experimental airplane built almost entirely of plastics is said to have been flown successfully in England before the Second World War! The substitution of synthetic fibers for cotton and wool has been due, not so much to shortages of these natural products as to certain properties of the synthetic substitutes which make them more desirable or at least more practical than the natural fibers. In many cases, similarly, the replacement of steel and wood by aluminum and plastics has been dictated by the superior qualities of these substitutes for certain functions. The Age of Substitutability has begun as the Age of Aluminum and Plastics.

Technological solutions could probably be worked out for most resource problems, but questions of value will increasingly conflict with more extensive use of both natural resources and synthetic substitutes. There are still large deposits of rich copper ores in the U.S. but exploitation of them would require open mines in the Cascades National Park and thus grossly damage a wonderful wilderness area. Similarly, titanium could be extracted from the sands of Cape Cod and various metals, including uranium, from the granite of the White Mountains, but not without spoiling the charm of these regions. The mass production of aluminum and of synthetic plastics implies not only large expenditures of energy, but also ecological disturbances and all too often the littering of landscapes and waterscapes with nonbiodegradable materials. Most resource problems therefore imply value judgments as to the comparative importance of economic factors and environmental quality.

Thus, the "limits to growth" are likely to be revealed not only by computer models of pollution, consumption, supplies of raw materials, and population growths, but probably even more by social choices concerning environmental quality and ways of life. Scientific societies know, or can learn, how to solve most of the material problems of life, but they do not know how to deal with the dilemma posed by the conflicts between technical solutions and humanistic, cultural values. This is apparent even in the formulation of policies dealing with the problems of wastes and recycling.

Wastes are an inescapable component of living and their disposal has affected many aspects of past and present societies. It provides archeologists with information concerning ancient ways of life; clinicians with opportunities for disease control; conservationists with an opportunity to recover otherwise wasted resources; technologists with the possibility of financial profits; esthetes with a justification for protests against modern ways of life; ecologists with environmental headaches. Junkyards of solid wastes are hallmarks of industrial civilization, especially in the United States, but wastefulness is not peculiar to our times or even to the human species. Many living creatures are wasteful and careless whenever they can obtain what they need without much effort and have more than they can use. Nature has its own junkyards.

The human traits that have resulted in today's consumption society have their origin deep in our evolutionary past. Like us, large apes in the wild are wasteful of food and soil their habitats. Old Stone Age hunters often killed many more animals than they needed for food, and the sites they occupied have been found to be littered with stone artifacts, animal bones and food residues. At the Olduvai site in East Africa, a low semicircular wall that probably served as a windscreen for our precursors one and a half million years ago was found to be surrounded by a vast assortment of animal bones and stone tools that had probably been tossed over the wall. Archeological sites of later periods also contain human artifacts allowed to accumulate for many generations. Items of stone, ivory, pottery or basketry are the Stone Age equivalents of the aluminum cans, plastic containers, automobile tires and other gadgets that litter our landscapes and waterscapes. The science of paleontology is largely built out of the solid wastes casually abandoned by ancient people.

The farming villages that emerged after the neolithic agricultural revolution seem to have accumulated less solid wastes than did the paleolithic sites, a difference that probably persisted in agricultural villages until recent times. But the relative tidiness of villages was due chiefly to the fact that the life of peasants was more parsimonious than that of hunter-gatherers. Poverty precludes wastes. Whereas paleolithic hunter-gatherers made new stone weapons and tools as they needed them and abandoned them casually, people in an agricultural economy owned but a few utensils which they used with care in order to pass them on to the next generation. Meat has long been a luxury for European peasants and bones were used in the local "pot au feu" or fed to household animals. European forests were extremely neat until recent times chiefly because dead wood constituted the only fuel of country people.

The garbage abandoned in urban areas of the past was largely organic material and therefore was decomposed by microbial action or consumed by animals. Pigs still roamed Broadway in New York City until the nineteenth century, feeding on the city's garbage, but the waste situation was nevertheless already bad at the end of the century, as revealed by the following statements published in the January 1887 issue of *Scientific American.* "The habits of the present generation are such as to give rise to more refuse matter and poisonous products than those of previous ages. The fuel we use, the articles we manufacture and the sewage combine to create more impurities than were known to our forefathers . . . [the best solution] would be disinfecting all animal matter by dry earth, and never allowing it to pollute our waters."

Since carelessness and wastefulness are innate in the human species, they manifest themselves wherever industrial societies produce more material goods than they really need. In reality, we are not more wasteful or careless than were people of the past but the nature of our wastes has changed and their amounts have increased. We would soon be buried under mountains of garbage if we did not burn it or cart it away as far as possible from human settlements, to dump it into water or bury it in landfills. But there are limits to disposal by these barbarous methods and the waste problem has reached a state of crisis in all industrial societies.

The need for new methods of disposal has called attention to the fact that most wastes contain valuable materials and should be regarded as wasted resources, not to be thrown away but to be recycled.

In nature, organic wastes are destroyed by microbes that decompose them, step by step, into their elementary constituents which can once more reenter the cycles of life. Wood, paper, and most other natural wastes are thus recycled by biological mechanisms, but these become inefficient or fail to function altogether if the wastes are too concentrated or if the conditions are grossly unnatural as is usually the case in the industrial and urban environments.

There are no biological mechanisms, furthermore, to deal with aluminum, steel and a host of synthetics with which nature has had no experience since they did not exist in the evolutionary past. Many different kinds of microbes can decompose natural fibers such as cotton or wool but not artificial ones such as nylon. Containers made of wood or cardboard are biodegradable, but not those made of aluminum or of certain plastics. New unnatural ways must therefore be worked out to deal with the wastes of the modern age. Under natural conditions, furthermore, wastes are usually diluted by the action of air or water currents, or by being spread over large areas of land or water, whereas they are first concentrated in modern techniques of disposal. Nothing could be further removed from the ways of nature than compacting solid wastes, then shredding them mechanically, and finally separating their constituents by sophisticated physicochemical procedures. Recycling technologies require initial steps of assembling and sorting out the wastes.

Precious metals obviously present the simplest situation since objects of gold, silver or platinum are not carelessly discarded. Indeed, they are commonly made of metal first mined and refined decades or centuries ago and that will find its way into other future objects or uses. The scarcity and economic value of precious metals make it inevitable that the largest percentage of them are constantly being recycled.

Nonferrous metals such as copper, lead, zinc, aluminum, tin and others that resist corrosion can be recovered and reused provided their state of dispersion does not make the retrieval so costly that recycling is economically unacceptable. Iron is the metal recoverable in largest amounts, usually in the form of steel from discarded equipment; large amounts of scrap are available in densely populated areas with heavy industries.

The decision to recycle or not to recycle is thus conditioned by a multiplicity of economic and social factors. During the war, collecting stations were set up to facilitate recovery of tin cans because this metal had to be imported and was then in short supply. Recycling of tin cans is no longer economically justifiable today, but the collection of aluminum cans may be more profitable because the production of this metal requires large expenditures of energy.

The value of the material recovered and the net energy gain are thus two of the essential factors in determining the social desirability of recycling, but the necessity to dispose of wastes is another significant factor. The present practice of burying wastes in landfills, dumping them into bodies of water, or burning them cannot be continued much longer and will in any case become increasingly expensive as garbage and sewage have to be moved further and further away from populated areas. Since the cost of disposal should be deducted from the cost of recycling, the latter practice will become more and more economically justifiable.

Two types of technology—admittedly rather simple—that are within my professional competence will illustrate how modern industries can be less polluting and less wasteful than were those of the past. Shortly after my retirement from Rockefeller University I was asked by a large brewing company to act as consultant on the environmental problems caused by the production of beer. The wastes from fermentation in the brewing processes are rich in organic matter as well as in other chemicals and grossly pollute the streams or lakes into which they are discharged. As the company in question wanted to build a huge brewery close to a nature reserve with a rich aquatic habitat, it was eager to organize the brewing operations and to design the brewery in such a manner as to avoid water pollution. This was achieved, not only by careful engineering design, but also by converting within the plant itself all the wastes into useful products such as fertilizers and animal feed. Even the carbon dioxide emitted during fermentation was collected for use. The brewery has now been in operation for several years and the nature reserve is as flourishing as ever.

I have also had some experience with the environmental problems created by lumber companies. Some phases of the lumber industry use large amounts of chemicals, sulfur being one of them, which used to find their way into the effluents and to destroy animal and plant life in streams or lakes. This is no longer permitted in many places and there is evidence that the effluents of the lumber industry can now be recycled or made essentially innocuous. Furthermore, the lumber industry naturally generates immense amounts of wood shavings, sawdust and other debris which used to be dumped in remote places, including canyons in the wilderness. These wood wastes are now used, in one form or another, as sources of energy for the lumbering operations.

In the final analysis, however, the best way to deal with the problem of wastes is to decrease the amounts being produced, for example by reasonable changes in the ways of life and especially by improving the durability of manufactured goods. In terms of energy costs, making a product more durable is always preferable to recycling. If the development of recycling techniques were to encourage obsolescence, the long-range effect would be a further loss in the art of producing goods of lasting value. Decreasing the amount of wastes produced per capita certainly contributes to the quality of civilization.

THE MERITS OF ENERGY SHORTAGES

Most discussions about energy are focused on shortages, costs, new sources and more efficient techniques of use—the general assumption being that the more energy we have and can afford to use, the better off we are. Like any other resource, however, energy is but a means for reaching a chosen goal or for achieving a desired end. If this can be done more effectively with less energy, so much the better. The increased energy consumption per capita which has occurred in most parts of the world has probably been beneficial to the majority of people, in the past, and would certainly facilitate life in underdeveloped countries now. But there are reasons for believing that the rates of consumption have reached a point of diminishing returns and may even be at a point of negative returns in most prosperous countries.

Until the nineteenth century, the only significant sources of energy were human and animal muscles, watermills, windmills, and wood. The Industrial Revolution began with these energy sources, which were renewable but limited in amounts. Wood, which fired the first steam engines and locomotives, became scarce in the United Kingdom as early as the beginning of the nineteenth century. Despite the inventiveness of technologists and scientists, the Industrial Revolution would not have gone far had coal not been readily available to replace wood. Coal supplies were at first thought to be virtually inexhaustible, and the belief that industrial growth could therefore go on forever was strengthened when petroleum and natural gas came into widespread use during the twentieth century.

Not only has the abundance of fossil fuels made possible the large-scale practical application of mechanical and chemical discoveries to industrial processes; it has also extended the range of materials that can be used in these processes. If some metals are in short supply, for example, energy can be used to extract them from lower-grade ores or to create substitutes by chemical synthesis. If pollutants are being produced by industrial processes, energy can be used to trap them and render them innocuous or to develop protective measures against their effects. The advances in agricultural practices and in the processing of food also imply large uses of energy. In the final analysis, much of our present wealth can be traced to the use of petroleum and of natural gas which has provided energy at a very low cost in convenient forms for more than a century. Only now are we beginning to realize how profoundly the massive use of irreplaceable fossil fuels has affected the countries of Western civilization, increasingly from decade to decade since the end of the nineteenth century.

The beneficial effects of high energy consumption are obvious: unprecedented levels of creature comfort, lengthened average life span, unprecedented mobility, more egalitarian opportunities for education, entertainment, and culture. These privileges have come to be looked upon as birthrights, with almost no awareness of the price that we are paying for our high-speed, high-rise, "hyped-up" civilization. Personal control of a 350 horsepower automobile is equivalent in energy terms to the power of an Egyptian pharaoh with 350 horses or 3,500 slaves at his command. The people of high-energy societies seem willing to pay almost any price to maintain control over this equivalent of horses and slaves, but it is highly probable that as the price of fuels goes up, they will have to make do with a somewhat smaller equipage.

Public concern over energy began in the 1960s when it came to be realized that both its production and use generate pollution and upset ecological systems. Then, in 1973, the Arab oil embargo, followed by sharp and repeated increases in the price of petroleum and natural gas, called attention to the fact that the supplies of fossil fuels not only are limited, but also are highly localized. An atmosphere of gloom then began to spread over the world because many people became convinced that the twilight of the energy era was inevitable and with it the decadence of industrial civilization. One of the main difficulties of our times comes from our failure to have realized that the so-called "Industrial Revolution" had been in reality not so much a scientific revolution as a series of great technological achievements made possible by the lavish use of cheap fossil fuels. Had this been understood sooner, it would have helped physical and social scientists to define better the forces at work in our society so as to prepare it for the phase-out of the petroleum era.

I spent my youth in French villages and small towns where electricity did not reach my home and therefore I appreciate the conveniences and diversified experiences that electric equipment contributes to modern life. For the same reason, however, I am well placed to know that ways of life involving small uses of energy are not necessarily miserable and brutish. I am even inclined to believe that some decrease in the levels of energy consumption that now prevail in many industrialized countries would result in better living conditions. A large percentage of the energy now used is simply being wasted and excessive use of it has a variety of undesirable effects on the structures and institutions of our societies and on ecological systems. Unfortunately, the situation seems to be worst in the United States from practically all these points of view.

With only six percent of the world's population, the United States uses some forty percent of the world's resources. These abstract figures acquire more concrete significance when it is realized that, for each single person in the United States, there is in use at any given time some 10,000 kilograms of steel, 160 kilograms of copper, 150 kilograms of lead, 125 kilograms of aluminum, 125 kilograms of zinc, and 20 kilograms of tin.

Admittedly, the relationship between energy consumption and the quality of life is so complex that comparisons between one country or one historical period and another are difficult to interpret. In the United States, for example, a significant percentage of the energy is used not for satisfying the needs of everyday life, but for producing foodstuffs and manufactured goods which are exported. Furthermore, the vastness of the country and the dispersal of human settlements result in much greater uses of energy for transportation than is the case for the more compact European countries. Finally, the United States has been favored until now with abundant local supplies of cheap energy—wood, coal, petroleum, natural gas—a situation which has made American people less energy-conscious than others who are less well endowed with fuels. As far back as a century ago, the energy consumption per capita was already much higher in North America than in Europe.

Granted that there are marked differences in energy consumption among industrialized countries, the massive use of petroleum and natural gas has had similar effects in all parts of the world that have been westernized.

- It has converted them from predominantly agricultural and village-centered social structures to technological and urban-centered societies.
- It has enormously increased the number of occupational specialties and created new problems for the coordination and control of the work of specialists.
- It has sharply reduced the participation of children in familial activities and thereby weakened familial institutions.

• It has rendered obsolete many of the former functions of villages, communities, and neighborhoods.

• It has made corporate and government bureaucracies the dominant institutions in the management of our lives.

• It has rendered societal management more complex and thereby increased hostility to all forms of public authority.

• It has made all high-energy societies much more vulnerable to various types of social breakdown.

These effects were greatly accelerated and exacerbated during and after World War II, a period in which the consumption of oil and gas skyrocketed with the upsurge of automobile and truck transport, with the suburbanization of the middle classes, and with farm mechanization which has driven immense numbers of tenant farmers to the cities in search of jobs.

Opinions differ greatly as to the dangers and merits of the social changes brought about by the massive use of energy but there is somewhat better agreement concerning its effects on natural and agricultural ecosystems.

While the combustion of fossil fuels has long been known to produce a variety of pollutants, the universality and extent of their effects are only now being recognized. For example, the acids produced by the oxidation of sulfur and nitrogen in internal combustion engines and in power plants are carried by air currents over vast areas and reach the earth's surface and the bodies of water in the form of acidic rains. These cause leaching of certain soil constituents, damage the vegetation, and alter aquatic life. Such phenomena were first recognized in Scandinavia and were thought to be caused in large part by acid rains originating from industries in the United Kingdom and Germany. Similar phenomena have now been observed in the lakes and forests of Canada, the Adirondacks and the Atlantic Coast. It has been calculated that if the present concentration of acids in the rain that falls over New England were to be maintained for ten years, the productivity of agriculture and of forestry in this region would decrease by some ten percent—a loss of photosynthesis, which, for this region alone, would correspond to the energy output of fifteen 1,000-megawatt power plants. In addition to acid rains, an immense diversity of pollutants originating from land masses are threatening many forms of ocean life. Reduction of photosynthesis in ocean systems would have disastrous consequences for global ecology.

Techniques might be developed to reduce chemical air pollution to a tolerable level. But there is no possible way to avoid heat pollution, because it is an inevitable consequence of both the production and consumption of energy. Even the so-called "solar" sources of energy—from radiation, wind, waterfalls, tides, or waves—are not as safe as commonly believed. While solar sources do not add to the total heat load of the planet, they may cause ecological disturbances by changing the distribution of heat. Any form of energy used on a large scale will disturb the natural patterns of energy flow through the global system. Whatever the nature of the power plants established on both sides of the North Atlantic, for example, they may soon be so numerous as to discharge into the Gulf Stream amounts of heat that will affect the marginal, ice-covered, subpolar regions, and start a process that eventually could result in the melting of the polar ice cap.

It is widely believed, although neither convincingly proven nor fully understood, that the consumption of fossil fuels can alter the climate by increasing the atmospheric concentration of carbon dioxide and of fine particulate matter. Carbon dioxide and particulate matter probably have opposite effects on the accumulation of heat on Earth, but not enough is known concerning their relative magnitudes to predict the climatic changes likely to result from the present rates of energy consumption. Experts generally agree, however, that at our current level of fossil-fuel consumption, *global* climatic disturbances can be expected around the year 2000 and *regional* disturbances will probably become significant much sooner. Another doubling of energy consumption in the United States would probably result in local disasters and a worldwide doubling would certainly upset the global ecosystem.

Clinical experience, epidemiological surveys, and experiments with animals concur in showing that longevity and health usually benefit from rather frugal diets and vigorous physical exercise throughout the life span. The people of many poor countries engage in large amounts of physical work during their daily lives yet can do it with less food than we usually consume and with a diet consisting chiefly of starch and vegetables, instead of sugar and meat. In certain regions which have a large percentage of very old people, many of whom remain physically and sexually active, food intake is relatively restricted and continuous physical work is the rule for both men and women. Many people in prosperous industrialized countries could thus decrease their use of energy and thereby improve their health by limiting their consumption of meat and sugar and by becoming less dependent on machines for work, transportation, leisure, and other occupations of daily life.

Mental health can probably also be undermined by excessive and unwise use of energy because this impoverishes our contacts with the external world. Any experience is likely to be weakened and distorted if it is passive; for example, watching nature through the windows of a motorcar or having the illusion of participating in the human encounter by watching a television screen. Energy from external sources can, of course, enlarge and diversify our contacts with the world, but all too often we tend to use it chiefly to minimize effort, thereby impoverishing the experience of reality. Our potentialities for intellectual performance, human relationships, or emotional experiences do not develop any better by viewing a television program than do our muscles while watching a sporting event.

In the past, the design of human settlements had to take into consideration climate, topography and other physical characteristics of the region. Such natural constraints resulted in a great diversity of architectural and planning styles, which accounted for much of the charm, interest, and also comfort of regional living conditions. The practical and esthetic quality of "architecture without architects" was a product of the necessity to cope with the natural environment.

In contrast, planners can now almost ignore the intensity of the sun, the cold winter temperatures, the impact of rain or snow on buildings, the necessity to adapt the steepness of the roof slopes to climatic conditions. The distance separating houses and the distance between home and work likewise are of little significance in modern planning. Instead of being concerned with local constraints, architects and planners now put their trust in the use of more and more energy to heat and air-condition buildings, to shelter people from stimuli, to move them from one place to another, and to bring utilities wherever needed. Avoidance of local constraints generates higher costs of building and higher consumption of energy. Most importantly, perhaps, it tends to decrease the esthetic diversity of architecture and the quality of human relationships. Landscapes are spotted with tacky houses, buildings become stereotyped, their occupants lose contact with other people and with the environment, communities disintegrate.

An isolated, free-standing house, surrounded by as much open ground as possible, has long been one of the ideals of American life. This ideal used to be compatible with the social and economic conditions of the past when there was much unoccupied, inexpensive land and when the family house was essentially self-sufficient—with its own woodlot, water supply from a well or a stream, food supply from the garden, domestic animals, and wildlife, and few problems about waste disposal. But conditions have changed. The free-standing house is now increasingly dependent on public services—for electricity, telephone, water, sewage systems, and dependent also on fuel for heating, transportation, road maintenance, snow removal, and practically all conveniences of modern life.

In the modern world, the way of life implied by the free-standing house involves such high social costs, especially with regard to labor and energy, that it may become an economic burden, too heavy for the average person and perhaps socially unacceptable.

The increase in energy costs may act as the catalyst for the design of buildings adapted to the natural environment and for a restructuring of human settlements perhaps based on greater clustering of habitations. This would make for economies in fuel consumption, in the maintenance of roads, in access to utility lines, sewage disposal systems, shops and schools; it would furthermore facilitate group activities and thereby foster community life.

The clustering of habitations in suburban and rural areas would also release land for agriculture, forestry, and even for re-creating semiwilderness areas. Many architects and social planners are now trying to design human settlements that provide both the technological advantages of cluster housing and the sense of privacy and space associated with the free-standing house.

Modern agriculture is increasingly dependent on multiple forms of industrial energy for the production and use of farm equipment, chemical fertilizers, insecticides and herbicides, irrigation and drainage. Scientific farming can thus be regarded as a complex technology for converting, so to speak, fossil fuels into crops that are further transformed into foods and other materials. Its successes are expressed in the phenomenal increase of agricultural production, and in the conversion of low-cost calories (fuels) into other calories such as food and fibers that are much more valuable for human beings. But farm technology has indirect costs.

The more agriculture depends on industrial energy, the smaller are its true yields, measured in terms of numbers of industrial calories needed to produce one unit of crop. For example, the thirty-four percent increase in food production achieved in the United States between 1951 and 1966 was accompanied by a 146 percent increase in the use of nitrates and a 300 percent increase in the use of pesticides. Even greater energy expenditure per unit of crop is likely as less fertile lands are put into production. Nor is there much hope that this situation can be improved, because, according to a recent report, the substitution of fossil-fuel power for human labor and the use of chemical fertilizers and pesticides have made present forms of farming just about as efficient as they are going to get. The production costs of agricultural products will therefore increase with the cost of energy.

It is likely, on the other hand that shortages of energy and its high cost will bring about beneficial changes in agricultural practices. The massive use of heavy equipment, of chemical fertilizers and of synthetic pesticides results in much ecological damage; soils become compacted and lose their humus; waterways are contaminated by erosion and chemical effluents; fixation of nitrogen by bacteria is reduced by nitrogen fertilizers. More emphasis on environmental and biological considerations, based on modern ecological knowledge, could lead to a decrease of energy use in agriculture and create scientifically the equivalent of the empirical practices through which the peasants of old maintained soil fertility generation after generation. It might have the additional merit of improving the picturesque quality of landscapes through better adaptation of agriculture to the geological, topographical and other natural characteristics of each particular region.

Different as they are, all the examples mentioned in this section have one aspect in common. In all of them, energy is used to decrease or eliminate the efforts required of the organism or of the system to continue functioning. A large percentage of energy consumption does not serve really creative activities, but only reduces the effort of adaptation to the challenges of the natural environment. This protective philosophy helps to make life easier, but in most cases it impoverishes the living experience. We inject energy into human and natural systems as a substitute for the adaptive responses that these systems would otherwise make. This practice tends to cause an atrophy of the mechanisms of response inherent in all living systems, thereby decreasing the formative effects of adaptation to environmental challenges.

Genes do not determine traits; they only govern the responses that organisms make to environmental stimuli. All organisms have potentialities that develop fully into functional attributes only when the need arises for their use. This is well recognized for the physical and mental attributes of human beings, but it is true also for microbial life. Just as muscles do not develop well if they are not used, so microbes capable of fixing atmospheric nitrogen will not do it if cultivated in a medium containing large amounts of nitrogenous substances that they can use for their metabolism.

The same general principle applies to social and ecological systems. Architects and planners tend to become less inventive when an abundance of energy enables them to ignore both the constraints and potentialities of the local environment. Natural mechanisms of adaptation, which could contribute to ecological diversity, to regional originality, and to soil fertility, may be inhibited by excessive energy consumption. Thus, one cannot evaluate the full effects of introducing high levels of energy into a system until one takes into account the extent to which these levels interfere with the adaptive and creative responses that the system would make under other conditions, more natural and probably more exacting.

The energy crisis will be a blessing if it compels us to develop healthier and richer ways of life by giving fuller expression to the adaptive and creative potentialities of natural systems and of the human organism. Trends in this direction can be seen in the present discussions concerning the comparative merits of centralizing or decentralizing energy production systems.

Strip mines, oil refineries, huge hydroelectric facilities, nuclear plants, high-voltage power lines are seen by many people as evidence of progress; but they represent for others threats to personal liberties and to the spirit of civilization. In contrast, all forms of solar energy—sunlight, wind, running water, the biomass—appeal to people who fear the garrison atmosphere associated with high-power technologies. There is, indeed, a fundamental difference between the social effects of energy derived from fossil fuels or nuclear reactors, and that derived on the other hand from the various manifestations of solar energy. The likely outcome of the former is social centralization and of the latter social decentralization.

Fossil fuels constitute highly concentrated forms of energy that can be readily shipped to any point on earth. Nuclear reactors generate enormous amounts of electricity wherever they are located. These two methods of energy production therefore lend themselves to technologic, economic, and social systems with a high degree of social centralization.

In contrast, few are the sites with continuous unclouded sunshine, with dependable strong winds, with large volumes of falling water, or with amounts of biomass sufficient for large scale operations. Since the various forms of solar energy are usually available only in small quantities at any given time in any given place, the first steps in their use must be carried out in fairly small industrial units, a necessity which favors social decentralization. Solar-derived energy sources, including the biomass, are thus best suited to social structures different from those based on the large sources of energy derived from fossil fuels or nuclear reactors.

At the present time, the great majority of persons certainly prefer to have abundant electricity on tap without giving thought to its origin, its environmental dangers and its indirect social costs. But a significant percentage of the public, which seems to be increasing, tends to prefer local technologies on a smaller scale, more compatible with social decentralization and cultural pluralism. The selection of energy sources will thus involve choices based not only on scientific considerations and cost-benefit analyses but also on judgements of value concerning the ideal form of society. The final outcome will probably be a complex mix of centralized and decentralized sources of energy—suited to the expression of the multiple facets of human life and selected to fit the conditions prevailing in a given part of world. It is almost certain, in fact, that we shall eventually live with two complementary kinds of energy, one designed for large-scale activities, the other compatible with the aspirations and tastes of small groups of people and with the genius of the place.

We are so conditioned to believe that the more energy we can afford to use, the better off we are that any thought of limiting its consumption creates in the general public a sense of gloom and even of panic. Yet, conserving energy should not be regarded as a last-resort policy with unpleasant effects acceptable only to avoid painful future shortages, but rather as a way to improve environmental quality and to enrich human life.

Using less in the way of resources and energy will unquestionably upset industrial and employment patterns, but it will also create new types of occupations by generating new habits and stimulating the production of substitutes for the present types of manufactured goods. More importantly, the necessity to change will stimulate the use of imagination in redesigning society to make it more human. The quality of life is determined less by the mineral and energy resources available to society than by the resources and the energy of the human mind.

CREATIVE ADAPTATIONS AND ASSOCIATIONS

On many pages of the preceding chapters, I have referred to the adaptive mechanisms by which human beings, though fundamentally the same in biological structure everywhere on earth, nevertheless have been able to use the invariants of the species *Homo sapiens sapiens* to live under a very wide range of conditions and to create an immense diversity of cultures and ways of life. The word "adaptation," however, has several different meanings because fitness can be achieved in many different ways.

On the one hand, adaptation can be brought about through Darwinian evolution and therefore result from specific changes in the genetic DNA molecules—a process that usually requires many generations. On the other hand, adaptation can be achieved much more rapidly both through physiological responses and through sociocultural manipulations that do not require any change in the genetic constitution. We humans are most likely to be comfortable and successful if we make conscious individual efforts to achieve physiological and social fitness to the places where we live, work and play, to the climate, to the food we eat, to the clothes we wear, and especially to the people with whom we have to deal. In present human life, the physiological and sociocultural mechanisms of adaptation are of far greater practical importance than the Darwinian genetic mechanisms.

There is more to adaptation, furthermore, than the achievement of fitness. In the majority of cases, the successful interplay between people and the physical and social environments in which they develop and function involves the emergence of attitudes, qualities and structures that amount to a true creative process. For example, we become adapted to dangerous environments or to difficult tasks by developing greater resistance or new skills. These creative effects of adaptive processes have been greatly neglected. As I consider them of extreme importance for the history of life, I shall open here a large parenthesis to present a few examples of creative adaptations as they occur in various natural forms, and even in systems which most people assume to be essentially lifeless.

Consider, for example, a handful of garden soil. The general assumption is that soil is made up exclusively of the inanimate, inorganic constituents of the earth except for the worms and insects it harbors. In reality, however, every grain of soil contains billions upon billions of various types of microbes. I know this from having been a soil microbiologist early in my scientific life during the 1920s and 1930s. In fact, it was my experience as a soil microbiologist which first made me look at the problems of fitness and adaptation from an ecological point of view. I learned, for example, that the kinds of microbes in a good rich garden soil would do poorly or die altogether in a sandy, acidic soil or in any soil placed under water. The adaptive relationships between any particular type of soil and its microbial life are extremely complex and are of great theoretical and practical importance because they involve the very creation of the earth's surface.

There would be no real soil on the surface of the earth if it were not for the presence of microbial life; there would be only the inanimate chemical constituents of the planet, as is now the case for the surface of Mars and the Moon. Without life, the surface of the earth would look as harsh and uninviting as that of other celestial bodies. The humus which converts the chemical constituents of the earth surface into fertile soil and covers the bedrock is produced by microbial life.

How it all began is a mystery but we know that humus is constantly being produced by the soil microbes as they decompose the dead bodies and products of plants, animals and other forms of life. The characteristics and amounts of humus in a particular place, furthermore, depend upon the chemical composition of the local earth's constituents and upon other local environmental factors which determine the kinds of microbes that grow in that particular place. Each type of soil, in other words, results from the creation of a system in which fitness is achieved between the total environment and its microbial population—a fitness which determines in turn what species of animals and plants are most successful on it in a particular climate. The genetic DNA determines the potentialities and constraints of each species living in and on the earth's surface, but all manifestations of life are the expressions of relationships which are conditioned, not by the DNA molecules, but by the creative interplay between microbes and their soil environment.

I shall illustrate these relationships with an example also taken from my own work as a medical scientist—an example which has in fact conditioned my interest in biological adaptation, especially in human life. In the late 1920s I worked as a microbiologist on lobar pneumonia in the hospital of the Rockefeller Institute for Medical Research. We knew that the pneumococci responsible for lobar pneumonia owe their ability to cause disease to the fact that they are covered with a mucous layer, a capsule that protects them against the natural defense mechanisms of the human body, and that this mucous layer is made up of a complex sugar that we called capsular polysaccharide.

There was then no known way of destroying the capsular polysaccharide except by treatment with strong acid which of course could not be used in the body. Being familiar with the potentialities of the soil microbial population and with its ability to decompose even the most esoteric kinds of substances, I postulated that there probably existed somewhere in nature a kind of microbe that could feed on the polysaccharide by digesting it with a particular enzyme. The word "enzyme" is the generic term for proteins produced by the body that enable it, in all living things including microbes, to utilize food. In 1929, I succeeded indeed in separating from the soil a certain type of microbe that could feed on the capsular polysaccharide and I separated from cultures of it the enzyme which enabled it to digest this substance. Animals infected with pneumococci could then be completely cured by injection of the enzyme which destroyed the capsular polysaccharide of the pneumococci in their bodies.

The anticapsular enzyme that I prepared in 1929 was the first antibiotic produced by a rational scientific method in the laboratory, but although highly active against pneumococcal infections in animals it was never tested in human beings for several practical reasons—one being the difficulty to produce it in a pure form on a large scale, another being the discovery in the early 1930s that sulfa drugs could be used against a great variety of bacterial diseases, including lobar pneumonia.

After having established the antiinfectious activity of the microbial enzyme, I tried to develop methods for its production on a large scale and this led me to an unexpected theoretical discovery that has influenced all my subsequent life. I had no difficulty in producing huge amounts of the soil microbe by cultivating it in a rich bouillon, but to my surprise and disappointment, the microbial mass thus obtained did not contain the anticapsular enzyme in which I was interested. I eventually recognized that this enzyme was produced only when the soil microbe was deprived of other nutrients and forced to feed on the capsular polysaccharide itself or on a related substance. The *adaptive* response to the necessity of using the polysaccharide as food was thus a creative one—namely the production of an enzyme. For good scientific reasons, biochemists and geneticists who worked later on related problems of enzyme production coined the phrase "induced enzyme," but for reasons of biological philosophy, I still prefer the phrase "adaptive enzyme."

I was naturally excited by my finding, in part because of its scientific originality, but even more because I immediately realized its relevance to other forms of life, and especially to human life. The fact that the enzyme was produced in a very short time as an adaptive response to a certain necessity proved that microscopic organisms possess potentialities that are expressed only under certain conditions and I postulated that this biological law was applicable to other forms of life including human life and human behavior. This was a mechanism of adaptation very different from that achieved by Darwinian genetic processes which, in higher forms of life, take place slowly, over many generations. I subsequently developed other examples of adaptive enzyme production and ever since that time—more than half a century ago!—I have been obsessed by the conviction that all of us are born with the potential capabilities for many different life-styles, but develop only those which can be evoked into activity by the proper conditions and for which we make the proper kind of effort, more often than not out of necessity.

In nature, creative adaptations can also commonly result from the fact that the various forms of life exist in intimate associations with other forms not genetically related to them. Such biological associations involve animals, plants and microbes in all sorts of combination; they depend for their survival on changes which make the associated species better adapted to each other, usually with creative effects. Indeed, many types of organisms long thought to be well-defined biological species have turned out to be associations of several different species. For example, the sea animal called the Portuguese man-of-war consists in reality of at least three different species that have become banded together in the course of evolution; one constitutes the float, a second the fishing tentacles that capture plankton, a third carries out the digestive functions. The various organisms constituting the Portuguese man-of-war are so interdependent that they do not live long after being separated from each other.

The word symbiosis was coined more than one century ago to designate the biological associations between algae and fungi that produce lichens, and it has now been extended to many other types of associations between genetically unrelated organisms. Etymologically, symbiosis simply means living together, but in practice it is now used almost exclusively in its initial historical sense, namely with the added meaning that each organism contributes to the welfare of its associates. One of the most extensively studied symbiotic relationships is the one by which legumes and several other plant species living in association with certain types of bacteria are capable of utilizing the nitrogen of the atmosphere. On the rootlets of these plants are swellings known as root nodules which result from the tissue response of the plant to the *Rhizobium* bacteria present in the root system. The enormous population of bacteria in the nodules derives its nourishment from the plant, but in return the presence of the bacteria results in the production of a hemoglobinlike substance which makes it possible for the symbiotic plant-bacteria association to convert atmospheric nitrogen into organic nitrogenous compounds that the plant can utilize for its growth.

Lichens exist in at least 20,000 varieties—on the surface of rocks, on tree trunks, even in the Arctic and Antarctic wastelands and in other inhospitable places which appear almost incompatible with life. Each variety of lichen is made up of two different microscopic organisms—an alga and a fungus—the complementary physiological attributes of which enable them to derive nourishment from situations where neither of the symbionts could live alone. The association of alga with fungus, furthermore, endows lichens with great resistance to many noxious conditions such as heat, cold or drought that would be fatal to most forms of life and indeed to the alga and the fungus themselves if they were not associated.

Lichens exhibit different forms and properties depending upon the particular species of algae and fungi involved, and upon the substances on which they grow. The so-called reindeer moss which is almost the only vegetation covering immense areas in the subpolar region is in reality a lichen, and so is the inappropriately named California Spanish moss which grows supported by trees on the western coast of the United States. An interesting aspect of lichen biology is that the adaptive association between two microscopic species, the alga and the fungus, results in the creation of much larger botanical structures that display vivid colors, have great morphological diversity, produce complex chemical substances and have other characteristics never exhibited by either the fungus or the alga growing alone.

Many other cases of creative symbiosis have been recognized during recent years. I shall mention only a few examples at random. Cockroaches cannot fully develop if deprived of the rickettsia normally present in specialized organs of their bodies. Ruminants owe their ability to digest the cellulose of grass to a complex microbial population present in their stomachs. Even we humans depend on certain bacterial species that we acquire from our mother at the time of birth for the maintenance of our intestine in healthy condition. Most surprising of all, infection with a certain virus can, under the proper conditions, produce in tulips the marvelous variegated designs that generated in Holland a kind of stock market speculation known as the tulipomania of the seventeenth and eighteenth centuries.

In certain cases, the symbiotic relationship between the associates has become so intimate that one of the components of the system is no longer able to survive alone. Any green plant, whether it be a gigantic tree, a cabbage or a water lily, is capable of using solar radiation for its growth through the process called photosynthesis. This feat is made possible by microscopic structures (organelles) known as chloroplasts which are located in the cells of the plants, and which are essential for the production of chlorophyl. Chloroplasts used to be regarded as just other constituents of the plant, as are the roots, the leaves, or the flowers. It has now been proven, however, that the chloroplasts and the plants that harbor them have different kinds of DNA, a fact which makes it almost certain that they evolved independently before becoming associated. Living tissues contain other kinds of organelles, called mitochondria, which play a crucial part in the production of the biochemical energy required for all aspects of life in animals and plants. Here again, the mitochondria have a DNA different from that of the cells in which they function and therefore are likely to have evolved independently from them before becoming essential to the life of the creatures in which they are found.

All attempts to cultivate chloroplasts or mitochondria outside the cells from which they are obtained have so far failed. These organelles seem, therefore, to have lost the ability for independent existence during the course of their symbiotic life. This makes it even more remarkable that the symbiotic adaptation of chloroplasts and mitochondria to cells genetically different from them has creative effects of such enormous importance as the production of chlorophyl or the utilization of biological energy sources.

A new and even more subtle phenomenon of biological association came to light in the 1940s when it was shown that some genes of a given bacterial type can be incorporated into another type and thus make the recipient acquire some hereditary characteristics of the donor. Gene transfer can be achieved by different techniques but the one commonly known as DNA Recombinant Technique (one of the procedures of genetic engineering) has been most extensively studied and applied. With this technique, genes from many different kinds of creatures—microbes, plants, animals and even human beings—have been incorporated into bacterial species which then acquire some properties of the creature from which the gene has been obtained—for example the ability to produce insulin or some other hormone. It was believed at first that gene transfer was only a laboratory artifact but it has now been proven that the phenomenon takes place spontaneously in nature. Gene transfer certainly plays a role in the adaptation of living things to their physical and biological environments.

Creative adaptations and associations may thus have been significant factors in the evolution of life on our planet. Darwinian evolution postulates that, in the competition for survival, the reward goes to the fittest. But the meek may also inherit the earth through the creativeness of their adaptations and associations.

It is commonplace to equate the earth with a spaceship, forever circling the sun, without pilot and without purpose, as would any inanimate object of its size and chemical constitution. When we travel over the earth, however, what we experience is the prodigious diversity of its landscapes and waterscapes, as well as of the living creatures it nurtures and of their ways of life.

I dislike the expression "Spaceship Earth" because it calls to mind a mechanical structure carrying a limited amount of fuel for a defined trip and with no possibility of significant change in design. The earth, in contrast, has many attributes of a living organism which is forever changing. It constantly converts solar energy into innumerable organic products and increases in biological complexity as it travels through space. This view has been developed since 1972 by the English chemist J. E. Lovelock who suggests that the surface of the earth behaves as a highly integrated organism capable of controlling not only its own constitution but also its environment. Lovelock used the name of the Greek earth goddess Gaia to symbolize this complex biological behavior of the earth.

Regarding the earth as a living organism is not a new notion. Otis

T. Mason, one of the early American environmentalists, wrote in 1892, "Whatever our theory of its origin, the earth may be discussed as a living, thinking being. . . ." The Gaia hypothesis is more concrete and can best be formulated in Lovelock's own words. "The physical and chemical condition of the surface of the Earth, of the atmosphere and of the oceans has been and is actively made fit and comfortable by the presence of life itself. This is in contrast to the conventional wisdom which held that life adapted to the planetary conditions as it and they evolved their separate ways." I shall restate the Gaia hypothesis in purely biological terms and supplement Lovelock's physicochemical arguments with concepts of creative adaptations.

The Gaia concept has its origin in the fact that the chemical composition of our atmosphere is profoundly different from what it would be if it were determined only by lifeless physicochemical forces. In an earth without life, purely physicochemical phenomena would produce for example an atmosphere containing approximately ninety-eight percent carbon dioxide with very little if any nitrogen or oxygen, whereas the figures for the chemicals present in our atmosphere are approximately 0.03 percent carbon dioxide, seventy-nine percent nitrogen and twenty-one percent oxygen. Lovelock presents numerous other examples of such profound departures from chemical equilibrium and postulates the existence of a global force which brings about and keeps fairly constant a highly improbable distribution of molecules. He believes that this hypothetical global force has its origin in the chemical activities of the countless forms of life which create and maintain states of disequilibrium through feedback systems.

Obviously, the fitness of an organism to its environment is an essential condition of its biological success and even survival. All living things seem to be endowed with a multiplicity of mechanisms that enable them to achieve fitness by undergoing adaptive changes in response to those of the environment. In higher species and especially in humankind, the biological mechanisms of adaptation are supplemented by adaptive social processes. Early in the present century, the Harvard physiologist L. J. Henderson pointed out that there was more to fitness than the adaptive potentialities of living things. As he claimed in his celebrated essay, "The Fitness of the Environment," fitness can be achieved only because the terrestrial environment has certain physicochemical characteristics that happen to be just right for life. This theory, however, now seems faulty or at least incomplete.

Certainly, present forms of life would be annihilated if the surface of the earth were very different from what it is now. They could not adapt rapidly enough if the salinity, the acidity, the relative proportions of gases, minerals and organic substances, or any of the other physicochemical characteristics of the earth were to deviate far from their present values for any length of time. In other words, the *present environment* does indeed exhibit fitness for the *present forms* of life, but—and this is what Henderson overlooked—the present physicochemical characteristics of the earth's surface, of its waters and of its atmosphere, would have been unsuited to the primitive organisms of the distant past. An atmosphere with twenty-one percent oxygen, for example, would almost certainly have been toxic for the earliest forms of life.

During the past 3.5 billion years, the global environment has progressively changed, probably as a consequence of the activities of living things, and living things also have undergone corresponding changes through a feedback process. The Gaia system postulated by Lovelock therefore seems to be a result of coevolution. Lovelock discussed several chemical examples of this creative process and stated: "The air we breathe can be thought of as like the fur of a cat and the shell of a snail, not living but made by living cells so as to protect them against an unfavorable environment."

A few years ago, theoretical discussions were instigated to consider the possibility of making the planet Mars suitable for human life—which meant providing Mars with water, oxygen, moderate temperatures, protection from ultraviolet radiation, and so on. The general conclusion was that Mars could be made habitable only through the progressive introduction of living species capable of creating, over an immense period of time, more and more complex ecosystems similar to the ones which have evolved on earth during more than three billion years. This analysis has helped us recognize the profound and innumerable changes that life had to bring about on the surface of primitive earth to create, for present living things, the fitness of the terrestrial environment that L. J. Henderson had taken for granted as the initial and normal state of affairs.

According to the Gaia hypothesis, the earth's biosphere, atmosphere, oceans and soil constitute a feedback or cybernetic system which results in an optimal physical and chemical environment and in the characteristics of living things. Repeatedly throughout his book, Lovelock refers to this equilibrium situation as "homeostasis"—a word coined half a century ago by the Harvard physiologist, Walter B. Cannon, to denote the remarkable state of constancy in which healthy living things maintain themselves despite changes in their environment.

The word "homeostasis," however, does not do full justice to the Gaia concept which implies, in addition, that living things have profoundly transformed the surface of the earth and its atmosphere while themselves undergoing continuous changes in a coevolutionary process. Practically all the examples that Lovelock discusses refer, in fact, to *creative* evolution rather than to homeostatic reactions. For example, the accumulation of oxygen in the air which became significant two billion years ago (as a result of biological photosynthetic activities), probably destroyed many forms of life for which this gas was poisonous. In Lovelock's words, however, "Ingenuity triumphed and the danger was overcome, not in the human way by restoring the old order, but in the flexible Gaian way by adapting to change and converting a murderous intruder into a powerful friend." Cybernetic mechanisms progressively brought about the emergence of biological species capable of living in the presence of oxygen and of using it for the biological production of energy. In this case, as in most other environmental changes, the Gaian way was not an automatic homeostatic reaction but a creative coevolutionary response.

The Gaian control seems to result in global homeostasis only over a period of time which is short on the evolutionary scale. One figure will suffice to illustrate the magnitude of the terrestrial changes that are continuously caused by life. In their aggregate, all the green plants now fix approximately 100 billion tons of carbon per year in the various forms of the biomass. About half of this conversion of solar energy into chemical energy by the vegetation takes place on land, the other half in the waters of the earth. It corresponds to more than ten times the amount of energy that all of humankind uses annually, even with its most extravagant technologies. Who can doubt that this continuous process of accumulation of organic matter and energy will continue to affect both the surface of the earth and the various forms of life. Furthermore, Lovelock himself points out that the process of change may pick up speed and complexity as a result of human interventions, and he appropriately quotes me in stating that, on a local level, profound coevolutionary changes have already occurred in certain terrestrial environments and in their biological systems during historical times.

In all parts of the earth with large human populations, for example, agricultural practices have brought about dramatic decreases in the numbers of animals and plants of most native species and have simultaneously increased the numbers of other animals and plants, whether of local origin or of imported species, produced for economic reasons. Agricultural practices have also profoundly and repeatedly transformed the physical structure of landscapes and waterscapes. Before human occupation, most of England was covered by a deep forest. Then deforestation made it acquire the physical and biological characteristics of the large open Saxon fields. As a result of the Enclosure Acts, these fields were progressively replaced by a patchwork of much smaller fields separated by hedgerows and ditches. In our times, however, most of the enclosures are being eliminated to permit the use of heavy agricultural equipment. Changes of a different nature, but just as profound, have occurred in many parts of the earth's surface as a consequence of human activities.

In the last chapter of his book, Lovelock explores the relevance of the Gaian hypothesis to the effects of human interventions into nature, and suggests that environmentalists often shoot at wrong targets because the resiliency of the earth, considered as an organism, probably makes ecosystems more resistant to pollution than commonly believed. In my recent book *The Wooing of Earth* I have dared suggest that some of the human interventions can even increase the biological creativity and diversity of the earth, as is apparent in many European agricultural landscapes or in the "mountains and waters" agricultural complexes of southern China. I shall conclude by expressing my wish that in the next edition of his book, Lovelock emphasize not only the homeostatic aspects of the Gaia hypothesis but also its creative aspects. This would be in the spirit of his statement that the Gaia concept is an alternative to the "depressing picture of our planet as a demented spaceship, forever travelling, driverless and purposeless, around an inner circle of the sun" whereas, in reality, earth is constantly changing through the agency of all the forms of life which are part of it, including humankind.

6

OPTIMISM, DESPITE IT ALL

CIVILIZATION AND CIVILITY

THE SEARCH FOR CERTAINTIES

HUMAN COMMUNITIES

WEALTH, TECHNOLOGY AND HAPPINESS

SOCIAL PRIORITIES

DAYDREAMING ABOUT THE FUTURE

6

OPTIMISM, DESPITE IT ALL

CIVILIZATION AND CIVILITY

It is somewhat reassuring to note that the word civilization was introduced for the first time during a period even more troubled than ours, shortly before the American and the French Revolutions. The Marquis de Mirabeau seems to have used it first in an essay entitled "L'Amy des Hommes ou Traité de la Population" published in Paris around 1757. In an unpublished essay, "L'Amy des Hommes ou Traité de la Civilisation," he gave credit to women for most of the improvements essential to what he regarded as civilized life.

As used by Mirabeau, however, the word "civilization" had a meaning more restricted than the one we give it now. For him, and for most philosophers of the Enlightenment, civilization denoted humane laws, limitations on war, a high level of purpose and conduct, gentle ways of life—in brief the qualities considered the highest expressions of humanness in the eighteenth century. Samuel Johnson refused to enter the new word in the 1772 edition of his dictionary because he felt that it did not convey anything more than the older English word "civility."

The meaning of the word "civilization" has changed with time, or rather it has come to include more and more diverse manifestations of human life—ranging from Greek rationalism to Venetian sensuality, from artistic expressions to scientific technology, from Jefferson's pastoralism to worldwide urbanization. A view of civilized life even more different from that held during the eighteenth century gained prominence with the successes of the Industrial Revolution. As new technological developments brought about spectacular increases in wealth, the level of civilization came to be expressed in economic terms such as the quantity and diversity of food available or of manufactured articles produced. There was justification for this emphasis on material values because agricultural and industrial advances made life more comfortable, healthier, longer and perhaps even richer in experiences for most people. It was also assumed that greater economic wealth would inevitably improve the spiritual quality of human life.

Even the most optimistic person now realizes, however, that although technological civilization has resulted in greater wealth and better health for many people, it has not increased happiness nor provided better conditions for harmonious human relationships—for what Samuel Johnson called civility. Mirabeau would certainly be surprised to learn that our chief criteria in deciding that a society is civilized are that it has moved its outhouses indoors, heats and cools its homes with electric power, owns more automobiles, washing machines, freezers, telephones, and other gadgets than it needs. Gentle behavior, humane laws, limitations on war, a high level of purpose and conduct and all the attitudes that Samuel Johnson grouped under the name "civility" are barely included in the criteria associated with the word "civilization." Arts and literature are still emphasized, but more for their entertainment value than as contributions to civility.

Humanists, many sociologists, and even not a few scientists have come to believe that scientific technology harbors a demon which is bent on destroying, if not humankind, at least many human qualities of life. For example, Jacques Ellul has reported as a *fait accompli* the takeover of society by what he calls "la technique," namely a set of forces that operate independently of human control. J. K. Galbraith also asserts that the modern technological society is an almost self-contained system responsive only to the direction of an essentially autonomous and anonymous "technostructure." Although the system still depends on the public, it secures acceptance of its products through an artificial demand created by private agencies and governmental policies.

Few, indeed, are the new products and processes of modern technology which are introduced to meet fundamental "needs" of humankind. Most of them appeal instead to the desire for change for change's sake; more often than not they satisfy "wants" artificially generated only on the basis of commercial criteria. But the artificial creation of wants has occurred throughout history and in my opinion is not necessarily bad. As members of the biological species *Homo sapiens* our essential *needs* are extremely limited and of no special interest, but as socialized human beings, we continuously develop new *wants,* some of which contribute to the growth of civilization. Elegant clothing, beautiful furniture, faithful recording of music are not biological needs of *Homo sapiens* but wants that make our species different from other animal species and enable it to enlarge continuously Mirabeau's view of civilization.

The chief dangers of technology come from our own proneness to let machines shape our lives. Henry Adams had perceived this trend when he visited the Palais des Machines at the Paris World's Fair in 1900. According to his account in *The Education of Henry Adams,* the biological and spiritual forces that had motivated human life in the past had become overpowered by steam and electricity; the cult of the dynamo had replaced the cult of the Virgin. In 1893, Chicago had a World's Fair in the classical Beaux Arts tradition, with hardly any reference to the industrial aspects of civilization that were already well developed in the United States; only European visitors remarked on the functional beauty of the modern tools and furniture on display. But the official mood had changed forty years later. The organizers of the 1933 Chicago World's Fair wanted to celebrate, not the classical civilization but the role of scientific technology in the modern world, taking it for granted that from now on machines would largely shape human life. The guidebook of the Fair proudly proclaimed: "Science discovers, genius invents, industry applies, and *man adapts himself to, or is molded by,* new things. . . . Individuals, groups, entire races of men *fall into step with* . . . science and industry [italics mine]."

A large sculptural group in the Hall of Science at the 1933 Chicago Fair was even more explicit than the guidebook in conveying the theme that machines were now essential to the welfare of humankind. The sculpture represented a man and a woman with hands outstretched as if in a state of fear or at least of ignorance. Between them stood a huge angular robot nearly twice their size and bending over them with an angular metallic arm thrown reassuringly around their bodies. The theme of the Fair was clearly that machines could now protect and guide human beings.

Thus, the dangers of technology do not come from social complexities that make it a Frankenstein monster independent of social control—as asserted by Ellul, Galbraith and many other contemporary sociologists—but rather from the fact that we accept technological imperatives instead of striving for other desirable human values.

There are many indications, fortunately, that the future will be shaped by more human concerns than was the recent past. An awareness is in the air that doing more and more of what we have already done, only bigger and faster, is not for the sane and that civilization can become an obscenity if it is technologically overdeveloped. We have also come to realize that because human activities are now so widespread, intensified and diversified, disasters on a global scale will certainly occur within a few decades if industrial societies continue to operate as they have during the past century. On this score, as stated in an earlier section, there are reasons for hoping that the future may not be as dark as it appeared during the 1960s and 1970s.

The world population is still growing, but not as fast as it was then. Industrialization continues to expand, but with technologies far less destructive than the old ones. The belching smokestacks, the torrents of polluting effluents will not be tolerated much longer in modern countries. Except in the case of nuclear warfare or of really catastrophic accidents, radioactivity is not likely to be a serious danger even if nuclear reactors become much more numerous than they are now. While it will never be possible to prevent completely the release of toxic substances, the dangers of pollution will probably be more limited and more rapidly detected because of better technical designs and also because wastes, instead of being casually discharged into the environment, will have to be considered as resources and used to some desirable purpose.

Great progress has been made toward repairing environmental damage. The resiliency of nature is much greater than was realized a decade ago and we are learning to utilize its normal recuperative powers. We can also correct environmental damage by artificial means as when mined and other degraded areas are reclaimed and made suitable for agricultural production or other activities. It is all but certain that this will become the universal practice in advanced industrial countries and that we shall not see again the kinds of environmental neglect that desecrated Appalachia early in the Industrial Revolution.

The *material* aspects of the future thus need not cause pessimism; they pose the kinds of problems that technological civilizations know how to solve, provided there is the social will. The *human* aspects of these problems, however, are much more elusive as had been perceived by John Ruskin who objected, not to industry itself, but rather to the loss of joy in work which he associated with factory work. Ruskin had rapidly perceived that the Industrial Revolution had brought about what he called the fragmentation of the human person; it had degraded and enslaved human beings by turning craftsmen proud of their handiwork into soulless operators of machines. It was this preoccupation that led him to turn from the criticism of painting to that of design in general, and to state in *Seven Lamps of Architecture* that the only right question to ask about a trade was simply, "Was it done with enjoyment—was the carver happy while he was about it?" As we shall see in the following and later sections of this book, our societies still create surroundings and ways of life in which many human beings, even among those belonging to favored economic classes, suffer from the experience of not belonging which is commonly expressed by the vague words "anomie" or "alienation"—denoting the attitude of persons who do not feel real parts of the groups in which they function because they do not share their values or norms.

THE SEARCH FOR CERTAINTIES

In 1978, the Rockefeller Foundation organized a meeting of physicians and medical scientists to evaluate the present health situation in the United States. All participants agreed that the national state of health had greatly improved during the past few decades and was probably now the best it had ever been: child mortality has greatly decreased, the expectancy of life is still increasing among men as well as among women of all ethnic groups; several infectious diseases have been practically eliminated or can be readily cured; much progress has been made in the management of chronic diseases that cannot yet be cured such as diabetes, hypertension, arthritis, pernicious anemia; contrary to general belief mortality is decreasing even in the cases of heart disease, stroke and many types of cancer. Despite these objective criteria of improvement in the control of disease, the participants in the conference acknowledged, however, a widespread feeling among people that the general state of health is deteriorating. This paradox was expressed in a phrase, "Doing better but feeling worse," which became the title of the book reporting the proceedings of the symposium.

Doing better but feeling worse applies to many aspects of life other than health in industrialized countries. Despite the general increase in wealth and in the satisfaction of material needs there seems to be a decrease in happiness as illustrated in the various protest movements of the 1960s and 1970s. The present mood of many young people on college campuses calls to mind a remark attributed to George Bernard Shaw as he observed the healthy and prosperous, yet disenchanted young adults of England during the 1930s. "They've got enough food, sexual freedom and indoor toilets. Why the deuce aren't they happy?" The answer to this question seems obvious for the college populations of the late 1970s. Many students, and young faculty as well, are unhappy because they are pessimistic about the prospects for jobs and about their economic future, but other, more important factors are responsible for the mood of despondency now so prevalent in the countries of Western civilization. Even among young people who are financially independent or who are assured of a lucrative job, many are those who often feel like shouting, "Stop the earth, I want to get off." Their attitude is not one of panic or of violent hostility to present conditions, simply one of lassitude and disenchantment for which it might be useful to create, if it does not exist, the word "blah-ness."

We live in a difficult period, but this has happened many times in the past. There is nothing new in the belief that one has been born in a calamitous or dull period of history. Consider, for example, the mood of the person who wrote, "The world has grown old and has lost its former vigor . . . the mountains are gutted and give less marble, the mines are exhausted and give less silver and gold . . . the fields lack farmers, the sea sailors." These words fit the mood of contemporary doomsayers but in fact they were written during the third century A.D. by Saint Cyprian at a time, some 1700 years ago, when the people of the Roman Empire were losing faith in the structure of their society. Another dark period was the late tenth century when the Nordic invasions and a series of natural disasters led many people in western Europe to believe that the year 1000 would see the end of the world.

Barbara Tuchman has recently shown in *A Distant Mirror: The Calamitous Fourteenth Century* that never was so much written about the *miseria* of human life as during the early part of the fourteenth century; this was the time of the Black Plague and people lived in daily dread not only of pestilence but also of famine, insurrections and wars.

While the Renaissance is commonly assumed to be a period of unbridled faith in the human condition, its very triumphs caused many scholars of the time to be alarmed about the future. In 1575 the French jurist and philosopher Louis LeRoy, a man reputed so wise that he was considered the modern Plato, published a book in which he expressed his worries about the destructive effects of the new knowledge and the new inventions. His book, *De la Vicissitude ou varieté des choses en l'univers,* immediately became popular all over Europe, probably because its mood fitted the anxieties of the period and assuaged the "future shock" of the time. During the post-Napoleonic era, the various European countries expressed, each in its own way, disenchantment with the time and with the human condition and tried to escape from it all into the various forms of romanticism and bohemian life.

Periods of doubt and gloom, with symptoms similar to those we experience now, have thus repeatedly occurred in the past—indeed whenever and wherever the ways of life have been rapidly disturbed by spurts in general knowledge or by social and technological innovations. Phrases such as "the times are out of joint," "all coherence is gone," "Weltschmerz" and "le mal du siècle" have passed into the language and make it clear that rare have been the idyllic phases of civilization in the past. Similar phrases have been coined to convey various forms of gloom in our own century.

Auden's poem "The Age of Anxiety," published in 1947, reveals the extent to which postwar life was haunted by ghosts of undefined but devastating guilt and by a sense of universal shame. Writing in *The New York Times* and the *Wall Street Journal,* James Reston introduced in 1967 the phrase "new pessimism" to denote the fears resulting from the widespread feeling that many problems of the modern world are caused by scientific technology, yet are not amenable to scientific control. The current phrase for the confusion of our times seems to be "The Age of Uncertainty" used by J. K. Galbraith as the title of a book based on his recent television programs. Until World War I, according to Galbraith, "aristocrats and capitalists had felt secure in their position, and even socialists felt sure in their faith," but these age-old certainties have been shaken by the two world wars and by the Great Depression.

Answering to a criticism by a Buddhist leader that his book *The Age of Uncertainty* lacks a central guiding principle, Galbraith recently stated that there could hardly be such a principle because life is a stream that flows without guidance. In this light, according to him, the only valid approach to improvement is to modify social systems stepwise, more or less empirically, in the hope of moving them in the direction of greater harmony between people, greater physical well-being and thereby greater happiness. If these efforts fail, they can be redirected on the basis of experience.

Empirical approaches to social problems are limited by a fundamental difficulty. In the absence of a central guiding principle, human activities tend to become ends in themselves; they proceed on their own course and become increasingly unrelated to general human concerns. Yet, most people are not satisfied with purely material well-being; they long for permanent values, however vaguely. Even while enjoying ephemeral and trivial satisfactions, they want their lives to be organized around lasting certainties that they consider important. As Galbraith himself stated, "I see no purpose . . . in being an economist as such. It's a very dull way of life. The only purpose of being an economist is to see if you can add somewhat to the current of happiness." A central guiding principle is needed in all human activities.

The uncertainties Galbraith refers to in his book are chiefly in the domain of economics and politics, but various changes in social structures also contribute to the atmosphere of anxiety and uncertainty of the modern world. For example the family, whether in its nuclear or extended form, used to be responsible for the transfer of educational and ethical values as well as for many other aspects of socialization. These roles, however, are increasingly taken over by other social institutions with the result that there is a corresponding decrease in the certainties associated with familial ties and loyalties. None of our social institutions adequately fulfills the roles that used to be played by the family and the church.

All over the industrialized world, furthermore, political, economic and social organizations tend to become so large and complex that they can no longer be apprehended by the human mind, with the result that people feel like anonymous dispensable cogs in the social megamachine. Among the most prevalent symptoms of anxiety and uncertainty are the sense of helplessness in the face of events that appear beyond our understanding, and the feeling of loneliness that comes from the impersonal character of many social relationships. Modern science contributes still further to the social atmosphere of bewilderment by giving the impression that human behavior is governed less by rationality than by forces over which we have no control. One of the unfortunate effects of widespread but shallow scientific information is the feeling of many persons that it is unjustified to take pride in being human. According to them, the power of deterministic biological forces as well as the immensities of geological time and of cosmic space reduce human qualities to those of a puppet.

Having lost the former belief that the world was designed for human life, large numbers of highly educated people try to find intellectual and emotional substitutes for the lost certainties by accepting astrological or mystical doctrines assumed to be the expression of deep wisdom because they originate from ancient times and distant—preferably Far Eastern—places. The appeal of fringe social or religious groups, even if they require blind obedience to a leader, is that they help to recapture the warmth of close human relationships and also some form of certainty that provides peace of mind.

While our period has been called the Age of Uncertainty it has also been marked paradoxically by an *increase in certainties* with regard to certain fundamental aspects of life, in particular human rights and environmental quality. Moreover, while old certainties were based on classical value systems, they have gained new strength from modern scientific knowledge, as can be illustrated with a few examples.

In the past, rules of behavior used to be largely derived from the religious and ethical systems prevailing in a given community. Now, many of these rules are being reinforced by factual scientific information. The biblical teaching that all human beings are descendants of Adam and Eve, for example, is compatible with the scientific fact that all have the same fundamental genetic constitution. The campaign for civil rights, and more generally for fundamental human rights, thus acquires scientific justification from the knowledge that all human beings belong to one well-defined biological species.

Scientific knowledge has confirmed and extended the concept that *Homo sapiens* can become fully human only by functioning in a human society. All human beings, furthermore, are born with a wide range of potentialities, but these can become expressed only to the extent that their development is permitted and stimulated by the proper environmental conditions. A good environment thus implies not only conditions suitable for the growth and operations of the body but also for the development and expression of mental attributes. The need for education and occupation in a proper environment is therefore one of the inalienable human rights.

Our genetic endowment strongly suggests that the human species originated in a semitropical savanna and our physiological needs show that we are still today biologically adapted to these environmental conditions. Since the immense majority of human beings could not long survive in the areas where they have settled, they are compelled to create out of the natural environment the conditions and the resources needed for human existence. Environmentalists may find it intellectually or morally shocking that the biblical teachings give man dominion over nature, but in practice scientific knowledge leaves no doubt that the persistence of human life implies humanization of much of the earth. On the other hand, the teachings of all great religions agree with modern ecological knowledge in recommending that the earth not be exploited as a quarry but used in a spirit of stewardship, with concern for the long-range welfare of ecological systems.

Ever since the seventeenth century, value systems have been considered outside the range of investigative competence of science. It is now realized, however, that this has led to a body of knowledge that leaves out of consideration many modes of existence of greatest importance in human life—from love to hate, from hope to despair, from salvation to damnation. No reliable way has yet been found to apply the methods of natural sciences to these aspects of human concern. Despite its bold claims, sociobiology has not yet established many significant links between purely biological aspects of life and those that are peculiarly human. On the other hand, many scientific journals increasingly use the word "ought" in some of their articles. When applied to social concerns the word "ought" has significance only within a given system of values. In fact it implies a return to a hierarchy of values which gives primacy to the dignity of the person and to the management of societies.

One of the new certainties of our age is that science cannot be purely objective as used to be believed. In their selection of problems, in their approach to them, and in the applications of their findings, all scientists—unconsciously if not consciously—are influenced by considerations of relevance to systems of value. We are concerned not only with human lives but also with life as a principle, not only with the quality of our local environment but also with the earth as an ecosystem. We are beginning to move away from the abstract concept of society toward the more compelling moral values of human communities.

HUMAN COMMUNITIES

Everywhere on earth during prehistory and early historical times, most human beings lived in small groups—whether in stable villages or as wandering bands. In a recent study of the primitive Yonomamo people of Venezuela and Brazil the anthropologist Napoleon Chagnon found that when a village grows to more than one hundred persons, it begins to break apart in strife. Tensions develop also in Amish villages when their population exceeds five hundred. All over the earth, the typical human *community* has always consisted of at the most a few hundred people. Surprising as it may appear, this is still true today in the form of neighborhoods even in huge modern cities. Many novels and scientific studies have been written about the social and emotional aspects of community problems but chiefly in towns and cities. The probable reason for the neglect of the village by novelists and sociologists is that, while most human *life* has been spent in small communities, *civilization* has largely emerged from cities, small or large.

Most of the world-famous cities did not start as agricultural villages that happened to grow larger with time but rather as trade centers in which people coming from other parts of the world exchanged goods and information. Many cities may have begun as places where people assembled to worship a particular god or goddess, and where they developed and practiced some collective system of belief. From the beginning, cities were thus places that fostered various aspects of the human encounter which helped to make them not only seats of political power but also regional centers of civilization.

Cities have grown in size and importance throughout the historical period but their rate of development was greatly accelerated by the Industrial Revolution. The most obvious causes for this acceleration were the need for a large labor force in the factories of the mushrooming industrial centers; the ease with which food and other materials could be transported over long distances by the newly developed railroad system; the mechanization of agriculture which progressively reduced the numbers of workers needed on the farms; the ever-increasing need of clerical work and other services resulting from the greater complexity of administrative requirements within the business organizations themselves and in their relation to governmental agencies. To varying extents, all these social forces are still at work now and their effect on the migration from the country to the city is further increased by the worldwide prevalence of unemployment and the feeling of unemployed people that their chance of finding work and, if need be, welfare help is greater in large cities than in villages.

Because of these, and many other reasons, there were numerous world cities exceeding one million inhabitants in 1978. In the so-called developing countries, which really means poor countries of the world, several cities have reached a population of ten million people, thus surpassing New York and London. It is said that if Mexico City continues growing at the present rate its population will be of the order of fifty million early next century.

In most cases, it is difficult to give a valid figure for the population of the world megalopolises because many of them are not isolated cities but rather the center of areas with high population densities. For example, whatever the exact figure for New York City as an administrative unit, the more significant figure is that which includes the populations of those parts of New York State, New Jersey, and Connecticut which are adjacent to New York City proper. The same could be said for London, Paris, Tokyo, Shanghai and any other megalopolis. There is no doubt therefore that the whole world is in the process of becoming urbanized. However, it is not certain that, as widely predicted, urbanization will continue to increase. We may be witnessing one of the many historical trends which have been extended to the point of the absurd and are bound to end, if not in disaster, at least in some form of collapse.

There are many persons in fact who believe that cities are no longer essential, and can survive only if they are so designed and managed as to greatly enhance the quality of life. In this light, the highest priorities in urban management should be to increase not the efficiency of cities, but rather their amenities because in the long run, these may be the only factors that will justify their existence. If people are not enjoying the urban environments in which they live, they will move out to places where the communication revolution will enable them to have, through artificial means, the experiences and contacts that only urban life could provide in the past.

In any case, town planners have begun to envisage a time when we shall no longer want to spend fortunes on immense transportation systems for people to do work in offices or factories that could be done at home by taking advantage of electronic communication and machinery. City life however has subtle merits that transcend economic considerations.

At its best, village life is characterized by stability, peace, comfort—but with too little excitement for some lively spirits. In contrast, the city symbolizes the opportunity for multiple contacts with new persons, ideas and experiences—in other words, the likelihood of adventure with its rewards and also the dangers that this implies. It is conventional to sing the charms of village life but the possibility of adventure has always had greater appeal for a large percentage of human beings. Throughout history, whenever they have had a chance, people have voted with their feet by moving from the village to the city. I am one of them. Even though I loved the small agricultural villages in which I was raised, I have lived in large cities throughout my life and, now, I continue to live in mid-Manhattan after more than ten years retirement from my university position.

I doubt that people settle in urban agglomerations only because this is where the better jobs are to be found. Throughout my long life I have known many sons and daughters of successful farmers, in all parts of the world and especially in the United States, who could have become very prosperous on the farm yet elected the difficulties and inconveniences of city life—merely for the sake of adventure. Neither do people go to the city for its monuments, theaters, concert halls, rare trees, flowers, or rocks, however beautiful these may be. They go chiefly for the sake of the human encounter with its possibilities of social contacts, intellectual and emotional satisfactions—but above all, in hope of the unexpected. With few exceptions, it is only when these expectations are satisfied that people become concerned with physical amenities such as architecture or parks. In fact, urban amenities can be usefully defined only in terms of the contributions they make to the richness of the human encounter.

Everywhere in the world, people are more likely to gather where human activities are happening which either interest them, move them emotionally, or get them socially involved. The shops of artisans, tradespeople or booksellers used to be meeting places because they provided displays of the human enterprise. The popularity as a meeting place of railroad stations in the past, and of some airports now is accounted for by the fact that they also provide a vicarious sense of adventure through travel and through contacts with the outside world. Most construction sites and demolition scenes also draw an audience.

The situations which have the greatest appeal, however, are the ones in which the person can take an active part in public activities and therefore be actor as well as spectator. The piazzas, malls and open-air cafés are examples of such situations. For complex historical reasons, attractive public places outdoors are rather infrequent in the United States but increasingly during recent years the doorsteps of houses in Manhattan and the low walls surrounding fountains in front of skyscrapers are used as seats by strollers, workers, lovers and bums, even in the busiest streets. Watching the life of the city as it goes by provides endless entertainment for the spectator, who himself contributes to the spectacle by his attitudes and his remarks.

Natural and architectural splendors are of course assets to a city, but even more important are the spectacles that ordinary human events generate in the streets, the parks and other attractive public places in the open air. The human, as against economic, success of a city is measured by the opportunities its citizens and its visitors have to participate in its collective life. Cities differ widely from this point of view. In some cities, any activity in the street beyond washing the family car or watering the lawn is considered loitering. In others, the street is merely a superhighway leading to the next city. It has been said that the street life in some parts of Los Angeles "means bumping into people in parking lots outside supermarkets." In contrast, most world cities can boast of promenades like the walk in Paris along the bookstalls that line the banks of the Seine, in Barcelona the Ramblas lined with stalls selling flowers, birds and newspapers, in Rome Il Corso and indeed almost any street, in Manhattan Madison Avenue with its stylish people walking in and out of fancy shops or art galleries, or the Lower East Side with its noisy bargain hunters, its merchants and strollers of almost every ethnic group.

The biological limitations of the human brain make it difficult really to know more than a few hundred persons. Consequently the small village probably constitutes the most comfortable social unit of life, but this kind of comfort is bought at a price. The tribe or the village offers only a narrow range of human associations and furthermore tends to impose behavioral constraints which limit personal development. In contrast, the city offers greater freedom and a much wider range of choices. The variety of its places of work and of entertainment as well as of its specialized groups provides a broad spectrum of human activities and relationships from which to choose and to create one's own personal human neighborhood. Even more importantly, the streets, squares or malls, the cafés, restaurants and other public establishments provide the opportunity for chance encounters which may be extremely rewarding—precisely because they cannot be planned and therefore add unexpected components to life.

Since the chief merit of the city is to provide a broad spectrum of conditions and of people from which to create the proper circumstances to act out one's own way of life, the diversity of the urban environment is far more important than its efficiency or its beauty. The great cities of the world acquired a rich diversity from their complex historical past and this is one of their greatest assets. It may be exhausting and traumatic to live in New York, London, Paris or Rome, but cities are great to the extent that they offer a broad range of atmospheres and public spaces that each citizen can use as stages to act out his own life and thereby create his own self-selected persona, as I did in the Paris Jardins du Luxembourg.

While the great world cities have continued to grow in size and complexity until the 1960s, there are signs that some of the most famous ones are beginning to lose population. In innumerable surveys recently carried out in North America and in Europe a majority of persons expressed a preference for life in towns and cities of moderate size and even for village life. Census reports show indeed that, despite continued general population growth, the population of very large American and European cities is decreasing somewhat while that of towns and small cities is increasing. People may not yet be ready for a return to the village but, in the rich industrialized countries at least, they have once again begun to vote with their feet, this time against megalopolis.

A partial explanation for this trend is the tenseness of urban life but I doubt if this is as significant a factor as usually believed because other forms of tension exist in towns and villages. Perhaps more important is the fact that, beyond a certain size, cities become more and more difficult and costly to manage. Demographers and planners who speak of future cities with fifty million people seem to forget that providing food, water and other necessities for such human masses, and disposing of their wastes, present problems for which there is no obvious solution. Beyond a certain point, furthermore, the greater availability of artistic treasures and of cultural activities made possible by the greater size of the city hardly contributes to the interest of life for the average person. The saturation point is soon reached with regard to the numbers of concert halls, theaters, art and scientific museums as well as sporting events that urban people can take advantage of within an hour's travel. Many urban planners believe indeed that with a population of 500,000 or even 250,000 people a city can afford all the amenities of the modern world, such as a university, a hospital with complete staff and equipment including that needed for the most sophisticated brain surgery. A city of that size can also support more facilities for culture, entertainment and sporting events than the individual city dweller can possibly use.

In my opinion, however, larger cities have the advantage of providing a greater diversity of environments, occupations and especially people. The American architect Louis Kahn was prone to describe a city as a place where young persons could wander the streets until they found what they wanted to do for the rest of their lives. Kahn was thus extolling the street as the vehicle for discovering opportunities and doing so in a metaphorical as well as literal sense. This is essentially the process of my own self-discovery in the streets and the parks of Paris as I recounted earlier.

The experience of human diversity found in some large urban agglomerations can be enriching, as well as entertaining, not only for young people but at all stages of life. In Manhattan during the past few months, my wife and I have had professional contacts, which acquired a personal quality, with four young Oriental women of approximately the same age who have been educated in the United States. One is Thai and works in a biological laboratory; another is a Mongolian who works in television; a third is a Japanese who is in a publication firm; the fourth is a Chinese involved in academic sociology and is very much concerned with ancestor worship. Moreover, two of these Oriental women are Buddhists but of different sects, one is Presbyterian and the other Jewish. Such ethnic and cultural diversity might perhaps be experienced in Berlin, London, Paris, Rome and a very few other large cosmopolitan metropolises, but must be extremely rare in cities of 250,000 people even though these can afford the most sophisticated amenities of the modern world.

One of the most frequent criticisms leveled against huge modern cities is that, by fostering the development of specialists in thousands of occupations, they tend to impoverish the range of social human contacts. Specialists develop ways of communicating so different from those of other people that they are almost as incomprehensible as if they spoke an unknown foreign tongue. Specialization is now so deep and widespread that almost nobody is able, or even tries, to apprehend megalopolis as a whole. Furthermore, the economics of wage systems in large cities makes it practically impossible for children to have a useful social role. Instead of being productive assets, they become expensive burdens and may not provide the psychological fulfillment hoped for by their parents. At the other end of the age spectrum, social security and other pension schemes finance independent livelihood for aged people, and thereby tend to separate them from their children and grandchildren. Family linkages become progressively weaker. Reciprocity of responsibility, once the strongest cohesive force holding society together, is no longer learned in a family context.

Communities and neighborhoods are similar to extensions of families in their reciprocity of concern for one another and are also casualties of megalopolis. High-rise apartments and suburban sprawl have taken the neighborliness out of neighborhoods. Meeting one's neighbors in the small affairs of daily life used to encourage a sociability that is much diminished when each person is driving a motor car or pushing or being shoved in huge public transportation systems.

In the United States, some seventeen percent of the work force is employed by some level of government and twenty percent by the five hundred largest corporations. The bureaucracies of these two sets of institutions, which are concentrated in or near huge urban centers, dominate many aspects of life. The sheer complexity of urban agglomerations means that a low profile type of government cannot cope with their manifold problems. As the administrative structure becomes more and more visible and is perceived to function below the level of expectation, it becomes increasingly vulnerable to protest, sabotage and social breakdown. Massive and continued unemployment in the central city, for example, inevitably results in social disruption.

In brief, it is a widespread view that urbanization destroys our competence in homemaking and in neighborly relationships; dependence on transportation systems devalue human feet and commonly paralyze us in frustrating polluted immobility; over-institutionalized education for masses of young people stifles and crushes their ability to learn. Countless voices are now affirming that the major institutions of urban society have become counterproductive and rob us of precisely what they set out to offer. Dissatisfaction with modern life seems to be the greatest in the most urbanized countries which happen to have also the highest gross national product per capita. Many affluent members of these societies, especially among the young, claim that they reject the material values and institutions of the contemporary urban-industrial complex and turn for guidance to esoteric religions, witchcraft or astrology. Some have tried to live in communes with "simpler" life-styles. Most of the communal experiments started in the United States during the 1960s and 1970s have failed but the few that survive and prosper suggest that modern life can be quite successful in small human communities. In fact this has been proven by the history of modern Israel where the communal system of kibbutzim, which started seventy years ago, is still living and growing.

The kibbutz is a collective village in which both production and consumption are equally shared by all members. Adults eat in a communal dining room and receive all services from shared communal units, as well as equal budget allotments. There are approximately 230 kibbutzim, accounting for some 100,000 people who constitute 3.5 percent of the total Israeli population. The kibbutz movement is now growing at a rate of almost three percent a year, somewhat lower than that of the Israeli population, but high enough to assure an absolute growth. Many of its members are in the second or third generations.

In contrast with other communal experiments, the kibbutz has rejected regressionist tendencies. It is based on the village, but challenges the common belief that rural standards of living and of cultural activities cannot be as high as those of urban centers. The agricultural and industrial technologies of the kibbutz are among the best in the world, taking full advantage of the most sophisticated machines, computers, and organizational know-how. In the kibbutz, technological and economic achievements are identified with the collectivity and not with the individual person. Instead of depending on competition for its success, the kibbutz attempts to establish a creative atmosphere of cooperation; ideally one's colleagues in work are one's friends after work. It operates according to a system of direct democracy in which every member has the same political rights and shares in the various phases of decision-making. Work being an essential aspect of the kibbutz's economic effectiveness, most of its members have considerable leisure time, which they can spend at hobbies according to their inclinations and wishes.

People are free to enter the kibbutz from the outside if they comply with its requirements and they are also free to leave it if its ways of life appear objectionable to them, for physical or social reasons. As a consequence of this absolute freedom, the kibbutz population has a large turnover, but the rate of desertions is less than twenty-five percent in the second and third generations. This leaves within the kibbutz system three quarters of the people who have accepted its formula of socialization.

Completely "new towns" of various dimensions have been or are being created in industrialized countries, in order to prevent or at least slow down the further growth of megalopolis. The "new towns" movement as well as the phrase originated in Great Britain but it is widespread now in Europe. The phrase refers to cities planned for 50,000 to 250,000 inhabitants. New towns have been much criticized for the coldness of their physical atmosphere and their lack of human warmth, but this may not be a fair criticism. It takes a few generations for a town or a city to acquire the qualities that make for a rich human life and that contribute to civilization. A recent study shows that Harlow, which is the oldest English "new town," since it was created in 1955, has become a real home for people of all ages. The chief complaints seem to come from teenagers who find life in it so well ordered and peaceful that they have to go to the poor sections of London when they want to engage in social work.

Most new towns have been planned in such a way as to facilitate the rebirth of neighborhoods. For example, the whole urban area is commonly divided in multiple subunits for approximately a thousand inhabitants—each with the equivalent of a "village square," a café, a school and other community services. There are also local recreation grounds and larger areas of more or less natural landscapes and waterscapes surrounding the city as a whole. Furthermore attempts have been made almost everywhere to provide local employment for a large percentage of the inhabitants.

Certain planners are more ambitious and claim that the time has come to rethink human settlements in terms of real villages because modern communication technology now makes it possible to provide anywhere in the country most if not all of the social and cultural amenities that used to be the privilege of huge cities.

For example, plans are being developed to convert a private estate of some four hundred acres, in the central part of France, into an extremely modern village. The enterprise is a *non-profit international organization* financed by various private and public funds.

The estate consists of some forty acres of native forest, the rest being pasture and farmland of rather low fertility. There are operating farm buildings, and a large mansion (built in 1890) still in fairly good physical state.

Woodland, farmland, pastures and a fast-running river will enable the new village to be almost self-sufficient with regard to production of food and energy, as well as for various social and cultural activities.

The new village will consist of some one hundred twenty private houses designed according to ecological principles to save energy; they will be clustered so as to keep most of the land available for agricultural production. It is hoped and assumed by the planners that people of different persuasions and occupations including artists will elect to settle in the village.

The farm will be modernized and new types of agriculture developed on the basis of ecological considerations and of energy requirements.

The existing mansion will be restored and restructured so as to serve as hotel, with facilities for public entertainment, concerts and scholarly seminars.

As funds become available from the operation of the enterprise, they will be used for certain kinds of research, especially for more biological forms of agriculture and for the development of better techniques in production and use of renewable sources of energy.

In the spring of 1980 a small group of British planners revealed comprehensive plans for what they called "the village of the future" which would combine the most advanced aspects of microelectronic technology and of scientific food production. The Dartington Hall Trust in Devon, which has sponsored the project, presented it as a "marriage of the microelectronic and the ecological revolutions," the goal being to create a community as self-sufficient as possible, with emphasis on energy saving and on preserving the quality of the environment. At the present time, "the village of the future" exists only in the form of a model built on a scale of 1:250, but its promoters feel confident that it will be converted into real communities "built on the pillars of energy conservation, no unemployment and self-fulfillment."

According to the model, the settlement will be built around a central square and marketplace, free of automobile traffic. The square will be surrounded by public establishments ranging from the village day-care center, to a spa and health center, a school including facilities for adult education and a building that the planners call a "skills exchange center." There will also be a café-restaurant, a library and other traditional buildings.

An unusual feature of the settlement will be a center for communal resources, called "cottage office" in the model, that can be used by visitors as well as by all inhabitants. It will provide public access to computers, message and answering services, television, teletex and other communication links, as well as pooled secretarial and accounting services. The emphasis on microelectronics makes it possible to offer a wide range of educational facilities for all age groups, as well as many types of entertainment.

The design of the settlement emphasizes energy conservation. Houses and workshops will be tightly clustered around the central square and there will be a hostel for visitors so as to avoid the need for extra bedrooms in individual houses. Renewable energy sources will be used wherever possible, electricity being derived chiefly from aero- and hydrogenerators. As much food as possible will be produced around the periphery of the village, recycling being done on site to produce not only compost, but also clean water, methane and other fuels by biological fermentation. A community depot will house cars and trucks for people who need to travel.

A special building will accommodate water purification, fish farming and a large greenhouse. Attempts will be made in this building to mimic the recycling systems of nature and to develop a "heat bank" that will store heat in the summer and release it in the winter. Water is one of the most effective and convenient materials for storing heat as has been shown by John and Nancy Todd and their associates in the New Alchemy Institute on Cape Cod, where aquaculture tanks located inside a greenhouse pay back their cost in heat-storing capacity alone.

The English planners of "the village of the future" intend to use complex technologies only where these serve essential human needs. In contrast, low technologies will be preferred in activities that foster authentic human work and self-fulfillment. For example robots will carry out dull repetitive jobs such as storage or retrieval of information whereas low technologies will be used in food production, design of clothes and other household activities.

Small-scale industries will be based on a mixture of high and low technologies for the production of goods either to be used within the settlement or to be sold outside. Few people will have only one job, the goal being to enhance individual fulfillment by blurring the distinction between work and leisure, and by engaging people in different tasks at different times. In addition to their main job, most people will be able to tend farms and gardens in the greenbelt surrounding the settlement since, as already mentioned, repetitive and boring tasks will be practically eliminated through the use of labor-saving technologies.

As yet, "the village of the future" is only an idea incarnated in a small model, but the Town and Country Planning Association of Great Britain has already been offered five sites ranging from thirty-four to 150 acres that could accommodate populations from five hundred to 10,000 people. Opinions differ as to the proper size. Some planners feel that a site of three hundred to five hundred acres would be desirable because it would produce more income and enable the settlement to finance the complex facilities that are an integral part of the plan. Others favor a smaller community size, not exceeding 2000 people, because it would be more manageable and more friendly.

All persons involved in this project realize, of course, that the future can never be planned in detail and that the model of "the village of the future" is "an exercise in the imagination" rather than a blueprint for a real settlement. However, they can derive encouragement from the fact that some elements of it have been operating for several years in the experimental settlement called "the Ark"—established on Cape Cod a few years ago by the members of the New Alchemy Institute, who are planning a larger unit on Prince Edward Island in Canada.

The survival and continued success of the Israeli kibbutzim, and of a few of the communes created in the United States and in Europe during the 1960s, prove that life in small settlements is quite compatible with the needs and aspirations of human nature. This is not surprising in view of the fact that the sharing and cooperating ways of collective life have been basic factors in the evolution of the human species during most of its existence, since the Stone Age. The modernistic small "new towns" and "villages of the future" may not be successful at first, but the chances are fair that these settlements will improve with time and be accepted in the near future, as is already the case for Harlow.

On the other hand, the unique joys and benefits of very large cities are also human creations and many people will probably continue to prefer them to smaller towns or villages. As mentioned earlier, the kibbutz population has remained around 3.5 percent of the total Israeli population during the past seventy years. Most Israelis seem to prefer living in Jerusalem, Haifa, Tel Aviv or other large cities. The hostility to megalopolis does not seem to come therefore from a fundamental dislike of urban life per se, but rather from the irritation against the huge, complex and anonymous institutions which control the activities of urban dwellers and often prevent them from enjoying the most valuable and unique attribute of the large urban agglomerations—their phenomenal human diversity.

WEALTH, TECHNOLOGY AND HAPPINESS

All sensible persons know that the best experiences of life are free and depend only upon our direct perception of the world—as when we feel glad merely to be alive, to be the kind of person we are, to be amidst unspoiled scenery, to be doing what we like to do. Philosophers and moralists have affirmed in a thousand forms for thousands of years that joy is not in things but in us. Lao Tzu expressed this truth in the phrase, "He who knows he has enough is rich." Thoreau went even further when he wrote, "The opportunities of living are diminished in proportion as what are called 'the means' are increased." Most people, however, believe that science and technology contribute greatly to happiness by adding to material wealth.

A few years ago I landed in New York at the Kennedy airport in midafternoon on a hot humid day in August. The taxicab that was taking me home was soon caught in a traffic jam which gave the driver an opportunity to complain about his job and about the sad state of the world in general. Noting my foreign accent, he assumed that I was not familiar with the United States and he proceeded to enlighten me about American life. "You are probably surprised to see so much automobile traffic in midafternoon," he remarked as we stood still in the sultry air saturated with gasoline fumes. "The reason there are so many people on the road at this time of the day is that the hours of work are short in the U.S. and most people can afford to own a car." And mopping his brow he added forcefully, "In the U.S. we all live like kings." Then he went on to describe how the place in which he had been raised not far from where we stood on the highway used to be a nice country neighborhood but had now become a slum.

My taxi driver was right when he bragged about the prosperity of America which gave to the average citizen almost the wealth of a king but paradoxically he was also right when he complained about having to live under slum conditions. Like the average person in the United States, he probably used enough energy in his car and in the several kinds of power-driven equipment in his home to be the equivalent of hundreds of slaves working for him. In this sense he was like a king, but a king who had very little freedom as to where and how he could use his slaves. In any case, there is no evidence that real kings are particularly happy.

The great majority of people have probably always believed that material wealth contributes to happiness but it was chiefly after Francis Bacon and the philosophers of the Enlightenment that the betterment of human life was identified with economic and technological expansion. This view of progress was perhaps most strikingly and crudely expressed around 1760 by the French economist Mercier de la Rivière when he wrote, "The greatest happiness consists in the greatest possible abundance of objects suitable for our enjoyment, and in the greatest liberty to profit by them." Two centuries later in the United States, the members of the Paley Commission on Materials Policy had an even more exalted view of the contributions made by material wealth to human life. The preface to their report issued in 1952 affirmed that continuance of the trend toward greater technological advances is necessary not only for assuring happiness but also for the cultivation of spiritual values.

Economic expansion and even progress based on scientific technology have now begun to lose some of their appeal, at least among certain social groups in the industrialized countries. The results of many surveys indicate that an increasing number of people are adopting ways of life based on "voluntary simplicity." Early in 1978, for example, a town of 2500 people in Northern California went on a voluntary week-long total power blackout, switching to camping stoves and oil lamps. The initial purpose was to protest an increase in utility rates but the unexpected result was that, according to reports by many of the inhabitants, living under simpler conditions generated a higher level of happiness in the community.

Surveys have been carried out during the past two decades in the United States, in Western Europe and in Japan to determine the effects of technological progress and the steadily rising standard of living on human health and happiness—these words being used to convey physical welfare, contentment, peace of mind. The results, essentially the same in all wealthy industrial nations, reveal peculiar paradoxes. Many people believe that knowledge and the state of health have somewhat improved during the past fifty years but a very large majority of them feel that inner happiness and peace of mind have decreased. As mentioned earlier, "Doing better but feeling worse" was the phrase coined by the experts of the Rockefeller Foundation symposium to convey this impression with regard to the medical state of the country. President Richard Nixon had expressed a similar view in his first State of the Union message when he stated: "Never has a nation seemed to have had more and enjoy less."

Another paradox revealed in two different surveys was that the majority of people who felt that happiness is on the wane, nevertheless did not hanker for bygone, happier times. When asked "Would you rather have lived during the horse and buggy days instead of now?" or "Would you have preferred to live in the good old days rather than the present period?" only a small minority (twenty-five percent in the first case and fifteen percent in the second) answered yes. The immense majority of Americans and probably of people in other industrialized countries reject the idea of returning to old ways of life.

These paradoxes can probably be explained in part by the human tendency to romanticize the past. Every language has a phrase for a deep-rooted belief in the virtues of the gone-by years: "the good old days," "les bons vieux temps," "cualquiera tiempo passado fue mejor." As far back as the days of the Roman Empire there were many allusions to *laudatores temporis acti,* "the glorifiers of the past." Another contributing factor to these paradoxes in our times is the rising of expectations. Even if things get "better" *objectively* speaking, they do not get better fast enough to meet our *subjective* expectations. Progress thus causes dissatisfaction in many people by inflating their expectations faster than these can actually be met. The sense of frustration caused by rising expectations is especially acute and widespread in the United States because each generation of immigrants and of their descendants has taken it for granted that the conditions and circumstances of its children would be better than its own—indeed as they have been so far. The American philosopher Nicholas Rescher has recently concluded from these facts that "as concerns happiness, progress sets a self defeating cycle into action:

Improvement → Aroused Expectations → Disappointment." Technical progress, furthermore, often results in what he calls "negative benefits . . . the removal or diminution of something bad" as can be done by medical treatment and waste disposal rather than the addition of new pleasures. According to Nicholas Rescher, "We must bring ourselves to realize that it is a forlorn hope to expect that technological progress can make a major contribution to human happiness except through these 'negative benefits.' " In fact, many other reasons make it increasingly difficult to achieve a betterment of human life through further technological and economic expansion:

- Once a reasonable degree of affluence has been reached, further increase does not result in better health or more happiness.
- Many persons seem to be more interested in leisure and simpler ways of life than in the acquisition of more wealth.
- Practically all the advances in technology and in prosperity have brought in their train undesirable consequences.

Concern for the environmental dangers created by the expansion of technology reaches deeply into all classes, as illustrated by the results of a survey recently conducted among 3000 Japanese scientists who were asked to state what they regarded as the most desirable and likely technological developments for the twenty-first century. In view of Japan's mastery in electronics, this field could have been expected to head the list, but, in fact, video telephone was at the bottom, whereas techniques for air-pollution control were at the top.

Most technological successes and increases in material wealth do result in dilemmas concerning many social values. For example:

• Economic expansion will remain essential as long as many people live under substandard conditions, but we fear the environmental consequences of further technological growth.

• Large modern societies need complex rules of organization and control, but this brings about almost inevitably a reduction of civil liberties.

• Technological advances result in unemployment, but we now realize that a significant work role is essential to the individual's self-esteem.

• Justice demands a more equitable distribution of the earth's resources among all people of the world, but this would probably disturb the economies of industrialized nations.

These and other related problems experienced by all advanced industrial societies have led thoughtful people to question the wisdom of further economic expansion and to state that we are approaching the end of the two-hundred-year period during which the Industrial Revolution was identified with the betterment of life. It is far from certain, of course, that attitudes concerning the desirability of material wealth have changed as profoundly as would appear from complaints about the economic rat race and from the advocacy of simpler ways of life. The loss of zest for economic expansion is found chiefly among a small percentage of the upper and middle social classes. It is probably rare among the less favored classes of wealthy nations and absent among the immense majority of people in the developing nations. Furthermore, protests against modern life are as old as modern life itself, as appears in James Truslow Adams's introduction to the 1931 edition of *The Education of Henry Adams,* half a century ago!

> There are not a few signs today that in this America of ours, there is wide revolt against the direction that our life has taken. We are no longer sure that wealth will create a satisfying scale of values for us. There is . . . a questioning of all concepts, including those of failure and success. Against the whole rushing stream of contemporary life, the individual feels himself rather powerless.

The value systems of Western civilization and of the American public in particular are probably more stable than appears from the current sociological literature. Many people have admittedly become alarmed by environmental degradation and they realize more clearly than they did half a century ago that wealth does not necessarily make for happiness. But few are the people who reject the Western belief in the possibility of progress and even fewer are those who refuse to incorporate the products of modern technology into their daily lives. If a decrease in *per capita* consumption of certain goods and of energy does really occur, it will be because in many situations, lower consumption can improve the quality of life and *is* therefore the best form of progress. This becomes obvious, for example, in the use we make of tools and machines.

Tools were devised to facilitate and broaden the range of human activities, as extensions of the human body. Their wise use can increase and enrich our physical and mental contacts with nature, but now we tend to use tools and machines as substitutes for our bodies and minds rather than to enrich our perception of reality. The result is usually an impoverishment of sensations and also a decrease in our ability to perceive the charm and diversity of the world.

Consider, for example, our sensual experience of the seasons. We can of course see the spring flowers or the autumn foliage through the windows of an automobile but not fully appreciate their brilliance; furthermore, the fragrance of the vegetation and of the earth, the song of the birds and the rustle of the wind, are masked or at least distorted by the smell of gasoline and by the noise of the engine. The experience of the automobile rider is poorer in quality and content than that of the walker. Walking enables us to be immersed in the totality of nature and to perceive its most subtle qualities with all our senses. Viewing scenery on a television screen is no substitute for the direct experience of nature.

There is a real danger, I believe, that we shall become less and less able to perceive the subtle qualities of nature if we lose the habit of making the effort needed for complete perception. The old biological law, "Use it or lose it," was first used to refer to sexual activity, but its more general significance is that the attributes of body and mind that are not used tend to undergo progressive atrophy. Just as our muscles become weaker for lack of exercise, so is the case for our memory and for our ability to perceive the world sensually.

The widespread use of computers may, for example, eventually influence the way we use our minds especially if it plays an important role during the early phases of the educational process. By providing us with a mechanism capable of amplifying the brain's capacity for logical sequential thought, computers are nudging us toward placing more and more emphasis on thought processes that are rigid and mechanical and that depend less and less on intuition and sensitivity. Computerthink may tend to make the left hemisphere of the brain take command of our activities. To the extent that this occurs, the likely result will be a decrease in human activities associated with the right hemisphere of the brain, such as feelings, intuition and artistic perception. People who believe that the capacity for computerthink logical sequential thinking is the most distinctive aspect of human nature may be eventually displaced by robots because these can outdo the human mind in purely sequential operations.

In my opinion, however, there is little chance that the logical, sequential operations of computerthink can ever effectively take command for a significant length of time because most of us prize the products of imagination and intuition much more than those of logical thinking. One admires Descartes but is likely to love Pascal for believing, as most of us do, that the heart has reasons that pure reason cannot understand.

Yet the microelectronic revolution is here to stay and its effects on modern life will be extremely diverse and of increasing importance. I shall therefore consider now some of the human responses to this new technology, having in mind that they will be determined not only by technological feasibility and the profit motive, but also by societal goals and values.

In 1946, the world's first electronic computer was switched on at the Moore School of Engineering in Pennsylvania. Called ENIAC (Electric Numerical Integrator and Calculator), it occupied a large room, contained 18,000 vacuum tubes and consumed enough power to drive a locomotive. A computer with equivalent capabilities can be built today for less than 100 dollars; it can be operated on flashlight batteries and be small enough to fit into a coat pocket. This new microelectronic technology is based on the possibility of imprinting tens of thousands of electronic components and complex circuits on chips of silicon one-fourth the size of a postage stamp. In the words of a committee of the National Academy of Sciences, "the modern era of electronics has ushered in a second industrial revolution . . . its impact on society could be even greater than that of the original industrial revolution."

Although just a few years old, microelectronic technology has already had numerous practical applications, from the manufacture of digital watches, to the control of automobile engines, and to the use of thousands of robots in industrial plants. Its effects will be qualitatively different from those of the first Industrial Revolution—the goals of which were to enhance human physical capabilities, to harness more energy, to shape materials, to travel further and faster, to fabricate as many objects and substances as possible out of the raw material of the earth.

In contrast, microelectronics extend our mental capacities by increasing the ability to store and communicate information, to compute, to carry out logical operations and to control processes. The electronic computer does for the mind, in extending its range, what the bulldozer and the steam hammer do for the arm. Even though they are still in the process of being developed, microprocessors have already made it possible to devise and operate factories in which computer-controlled equipment carries out an entire process of production. It can be taken for granted that electronic business machines will soon bring about sweeping changes in the way offices are structured and managed.

The microelectronic revolution has gone so far in such a short time, and is still progressing so rapidly that its consequences are only now beginning to be evaluated. This second Industrial Revolution will affect practically all types of work involving the use of machines and will therefore create profound disturbances in patterns of employment. It will probably increase the economic gap between rich and poor countries since the technologies that it will make possible or facilitate will require sophisticated knowledge and equipment and will furthermore decrease the opportunities for labor-intensive industry. Finally, it may even affect profoundly the views we hold of ourselves, of the outer world and of our place in the order of things.

In the past, new technologies have always affected our ways of life. We have had computers for some thirty years and they have been used in many commercial and industrial operations but until now, they have not been available for personal use on a large scale. In contrast, microelectronics will increasingly bring the new information technology into many domains of individual experience and thus affect not only our individual activities but perhaps also our personalities.

All aspects of the new information technology place a high value on the ability to think in abstract rather than concrete terms. Since the creation of economic wealth will increasingly require the use of information technology, this will therefore give advantage to persons with talents for abstraction. In fact, the trend has already reached into very young age groups. A child molding an object or an animal with clay operates in the traditional concrete manner, but simple and inexpensive microelectronic equipment is now widely available to represent objects and animals on a screen, an activity which involves essentially abstract processes of thought and experience.

On the negative side, the shift toward abstraction may impoverish many aspects of human life. As more and more of our activities have to do with the intangible domains of information, it may become more difficult to be "grounded," a word now in use to denote concrete connections with one's family, friends, community and culture. People may become more alienated from the world around them as they experience it in the ordinary course of life and it is probable that this will lead to some form of alienation from oneself.

It has been predicted that microelectronics will increasingly make it possible for people to work at home. As we use electronic equipment for communication, education and entertainment, we may retreat into an ever more restricted world with less and less concern for other people and for the general welfare. Since we have evolved in highly "grounded" social groups and have consequently little ability to thrive and grow in isolation, a world of electronic anchorites and hermits will increase still further the present trend to "explain" the nature of things instead of experiencing them with our senses—a trend which is contributing to the impoverishment of life.

The opposite situation, however, is also possible and certainly more desirable. Microelectronics offers immense opportunities for connection, action and creation. It will make possible new kinds of social and political networks, and greatly facilitate and accelerate alliances around issues. It may also give us new abilities for pattern recognition and thus enable us to detect the existence of unsuspected links between the various components of the physical world, society and ourselves. This would help us to understand our place in the cosmic order of things and enrich human life both in understanding and experience. As has been the case for all the developments of the first Industrial Revolution, the most important aspects of the second Industrial Revolution will be not the scientific and technological ones but those involving human choices based on judgments of value.

SOCIAL PRIORITIES

So many simultaneous crises are occurring all over the world, and so many different ways are being tried to deal with them that people everywhere agree on the need to "reorder priorities," but this is where the agreement ends. In the United States, there is some consensus on what should *not* be done: for example, allow further environmental degradation, destroy more of the wilderness, extend much further the superhighway system, increase the concentration of executive and administrative power in Washington, depend exclusively on American military power to settle international problems. However, while it is easy to agree on criticisms of policies that have failed or that have proven overambitious, it is much more difficult to formulate social priorities that lead to new constructive projects.

Certain priorities seem at first sight clearly defined by the acute problems of our times, such as unemployment, inflation, decreasing industrial productivity, deficiencies in medical care, crime in the streets, environmental degradation, poor systems of education or of public transportation and all other obvious evils of contemporary life. More often than not, however, procedures formulated to deal with emergencies turn out to be of questionable wisdom in the long run because they are counterproductive and commonly generate new problems for the future. Priming the economy to encourage the automobile and airplane industries will help to relieve unemployment now, but will create new conditions of massive unemployment when petroleum is no longer available, or when more sensible means of transportation are developed—as well as a wide range of catastrophic environmental problems that may result from more extensive use of automobiles and airplanes. A nationalized health service might be useful to people who are now deprived of medical care but will probably decrease the quality of medical services for a large percentage of the population. A larger police force with more sophisticated weapons and techniques of surveillance may help to decrease crime, but may also become a threat to civil liberties. An elaborate and efficient system of public transportation will make commuting easier and perhaps more pleasant, but will increase urban sprawl as it is now doing in Paris.

There are profound differences of opinion, furthermore, concerning the very nature of the projects that deserve priority. A few decades ago the revolutionary government of Mexico gave high priority to the creation of a magnificent Museum of Anthropology in Mexico City, probably the best and most enjoyable in the world. But the museum was created at a time when an immense majority of Mexican people suffered from acute shortages in the necessities and conveniences of life—as many of them, indeed, still do. The statements made by Mexican public officials at the time of the dedication of the museum, and inscribed on its walls, indicate that priority was given to this project in order to help the different ethnic groups of modern Mexico to achieve a better sense of national identity and unity. The People's Republic of China also made at its beginnings priority decisions that can hardly be justified in immediate economic terms. For most people in continental China, the living standards were then among the lowest in the world. Yet, shortly after gaining control over the mainland, the Communist regime financed elaborate archeological projects that brought to light hitherto unknown artistic treasures going back to the neolithic period and extending to the fourteenth century. The display of these treasures in Europe and America has probably done more for the prestige of modern China than the production of nuclear weapons, the agricultural achievements of the communes, or the role played by barefoot doctors in health improvement. Similar questions could be raised concerning the motivations that determined the priorities for building the cathedrals of medieval Europe during a period when the cities in which they were located had less than 30,000 inhabitants, most of whom lived under primitive conditions.

Noneconomic criteria also influenced the formulation of priorities in European nations at the end of the Second World War. In Germany, for example, the city of Hamburg rebuilt its Opera House even before the rubble created by aerial bombing had been cleared from the streets. In Poland, the ancient sections of Warsaw were reconstituted as they had been known and loved before the war, though the country was then acutely deficient in housing facilities and faced an uncertain future.

The slogan "we must reorder our priorities" is thus empty of meaning because each nation, each ethnic group, each social class, indeed each person has a particular view of what is most important for a satisfying life. To discuss priorities from purely economic and political points of view, as is the general practice among American social reformers, is to overlook that many of the values that contribute most to the quality of the living experience are of an intangible nature. The health effects of air pollution in Manhattan may be less important than the difficulty of enjoying the sight of the stars on a cloudless night and the impossibility of ever seeing the Milky Way. Also important, at least for me, is the fact that air pollution prevents the growth of lichens on the tree trunks and boulders of Central Park. Happiness depends to a large extent on certain environmental and social values that are rarely given a high place among priorities.

People have always objected, for example, to the intensity of noise in large cities—Manhattan being one of the most objectionable places from this point of view. Noise control was among the problems considered by the group of reformers that began working for improvement in American cities around 1870. As one of them stated, however, "If man came first [in our considerations], noise abatement would be effective in a week. But the machine comes first and it is easier for the machine to make noise." The New York City Noise Abatement Commission was disbanded in 1932 because its recommendations were not implemented. Yet, noise is for me one of the worst aspects of urban pollution not only because of its deleterious effects on physical health but because it interferes with the perception of pleasant sounds, such as those of bells, and intrudes into our personal space, thus discouraging the human encounter.

Since the sense of priorities is such a personal affair, I shall limit myself to consideration of a few of them in which I have taken a personal interest.

Working for peace and particularly against nuclear warfare should obviously have the highest social priority but I do not see any way in which I can contribute to these problems and therefore only express my conviction that, despite all the talk about deterrence, nuclear weapons will be used and that no medical preparedness will be really effective.

Massive unemployment among young people seems to me the greatest social tragedy of peacetime. Although I do not know how to solve it, I can at least express opinions that differ somewhat from those that govern present policies. It is generally assumed that the plight of unemployed young people can be made bearable by welfare help, including various forms of entertainment, but this seems to me a mistaken approach to the problem. Human beings cannot function except as parts of structured social groups. If they do not become part of the normal society, because they are permanently unemployed, they will organize themselves in social groups of their own, as they are doing now, and this will inevitably lead to destructive social conflicts in the near future. Providing welfare money for food, shelter and entertainment is not a solution to the social problem of unemployment. The need is for some profound restructuring of our society or, as a temporary stopgap measure, programs something like the Civilian Conservation Corps adapted to the conditions of our times.

Thus, the resetting of priorities cannot be based on purely economic reasons. While money will continue to be used in one form or another for most social transactions, there is evidence even now that nonmaterial aspects of life are becoming increasingly important in the personal and even national calculations of success. It is nonsensical to include in calculation of the Gross National Product (GNP) such negative elements as the high costs of managing overcrowded communities, of controlling their various forms of pollution and maintaining large police forces. The Organization of European Communities (OECD) has already formulated an alternative formula, called the Net National Welfare Index that includes qualitative values in the calculation of real national wealth. Japanese authorities are also trying to develop a similar index for long-range planning. The formulation of such an index of collective well-being should be given high priority everywhere in the world.

There are many ways of defining and measuring wealth and growth other than by material consumption. Personal liberty, free time, the creative arts, the literacy level, the consumption of drugs, the incidence of crime, suicide, or illness, all these items and many others that readily come to mind should be ingredients of an index of well-being which would be much closer than the GNP to what people consider significant factors in the quality of life.

Attempts to reformulate priorities on noneconomic criteria go counter the dominant trends in industrial societies and it may prove difficult to give young people the education that would prepare them for ways of life in which community spirit and some measure of self-sufficiency are as important as is now the acquisition and accumulation of money. The solution will not come from simple changes in the traditional school curricula or in pedagogical methods, but will require instead a new philosophy of educational systems.

Self-sufficiency is an acquired trait. It can best be developed if the individual person experiences several changes of scene and of occupation, preferably during youth. In the past, this broad aspect of education generally took place in the family where the child was progressively incorporated into various social occupations and given increasing responsibilities. In our times, however, family life rarely provides the child with the variety of experiences and responsibilities that are required for individual development. Nor is the present school system organized in a manner that enables it to carry out satisfactorily the process of socialization that was the responsibility of the family in the past.

As presently constituted, schools are enclosed institutions in which it is difficult to prepare the child for the multiple and complex environments of adult life. They could fully complement the family as centers of learning only if they managed to incorporate into their systems such devices as classes linked to practical work experience or transferrable credits between institutions of very different nature. Methods of qualification would have to recognize, furthermore, kinds of competence to complement those evaluated by the traditional examinations and grades. Our present form of institutionalized learning was designed for a fairly stable society but will be increasingly counterproductive if, as is probable, we continue to live in rapidly evolving open societies.

Education and learning must not only be more dispersed throughout the society than they are now; they must also be spread more continuously throughout the life span in an attempt to keep pace with social and technological changes. The necessity to continue learning in order to remain competent may have the advantage of exposing middle-aged people to the humanities that they had scorned during their early school years, an exposure that may become more and more valuable in the modern world.

Technological societies know how to create material wealth, but their ultimate success will depend on their ability to formulate a postindustrial humanistic culture. The shift from obsession with quantitative growth to the search for a better life will not be possible without radical changes in attitudes. The Industrial Revolution placed a premium on the kind of intelligence best suited to the invention of manufactured articles, as well as to their production and distribution on a large scale. In contrast, a humanistic society would prize more highly skills facilitating better human relationships and more creative interplay between humankind, nature and technology. In future societies, the most valuable people might be, not those with the greatest ability to produce material goods, but rather those who have the gift to spread good will and happiness through empathy and understanding. Such a gift may be innate in part but could certainly be enhanced by experience and education.

Profound changes in the educational system will not be achieved by decrees from central authorities. They will require as many different experimental programs as possible, in the hope that the successful ones will serve as examples and eventually create a new consensus. Technological societies are so complex that they dread the risk of human error. For this reason they tend to reward those specialists who are least likely to make mistakes, whether they fly airplanes or run computers. They tend to discourage those who wish to become involved in really new ventures that might create risks and upset the sociotechnological apple-cart.

Occupational and intelligence tests measure chiefly dimensions of the human mind that favor safety over creativity. Yet, societies can adapt to new conditions and really progress only by encouraging experiments, and giving license to take risks—whether in technology, land use, health or education. In fact, this is the way nature proceeds to achieve adaptation and evolution. Nature is not efficient, it is redundant. It always does things in many different ways, a number of them awkward, rather than aiming first at perfect solutions. To improve human life instead of simply producing more goods, industrial societies will have to try many different ways of dealing with future situations instead of depending on the decisions of a few experts—because experts tend to be concerned primarily with means and efficiency rather than with goals and with the creative diversity which is essential for a richer life and the continued growth of civilization.

In many cases, the search for definite "solutions" will be replaced by "adaptive interventions" taking the form of progressive incremental modifications—indeed of "muddling through." Warren Johnson, professor on the San Diego campus of the University of California, recently defended in his book, *Muddling Toward Frugality*, the appealing thesis that the need to adapt to more frugal ways will provide us with the opportunity to develop new life-styles that will be happier because richer in personal experiences. For example, as I pointed out earlier, real shortages of energy might well be a blessing in the long run because they will encourage us to cultivate some of our physical and mental potentialities that remain undeveloped if we find substitutes for them in mechanical devices; more of us would become able to sing or to play a simple musical instrument if canned music were not so readily available. Shortages would also induce us to make the areas where we live more diversified and therefore more satisfying if it were not so easy to travel long distances in search of attractive surroundings.

The need for new policies of land use is another type of social priority which provides the chance for environmental improvement. Except for the real wilderness, the most appealing aspects of nature are usually found in a few public parks, but more generally in agricultural areas and large private estates. Under present conditions, however, farming is no longer economically viable in the vicinity of urban agglomerations and most large private estates are destined to be abandoned, not only because of taxation but also because they no longer fit the tastes of young generations. Once agricultural soil is no longer farmed or large estates no longer taken care of by their owners, the land is rapidly invaded by brush and loses much of its esthetic appeal. The magnitude of this aspect of land use in the vicinity of urban centers can be appreciated from the fact that there exist some 80,000 acres of open land within a thirty-mile radius of midtown Manhattan, thus providing the possibility of haphazard and generally hideous development of almost one million dwelling places. As similar conditions exist in many parts of the American continent, as well as some parts of Europe, new policies of land use have become social priorities in several advanced nations.

One can think of many possibilities of management for the large areas of open land that still exist near urban agglomerations: for example, allowing them to return progressively to a state approaching that of wilderness; establishing greenbelts; creating parklands designed for public use; developing human settlements combined with large open public spaces and corresponding for example to the garden city formulated by Ebenezer Howard in England, to the Broadacre ideal of Frank Lloyd Wright, or preferably to the "villages of the future" described in the preceding section. In most places it will probably be desirable and possible to reintroduce some form of agricultural production, especially of perishable crops, if only to make fresh and tasty fruits and vegetables once more available to the city dweller. It is likely that we shall end up with a mix of these different types of management but this cannot be achieved without generating controversial problems of zoning.

Zoning policies have generally aimed at achieving some form of socioeconomic and occupational segregation whereas they should increasingly incorporate considerations of ecology and of environmental perception. Instead of being based on segregation, the new philosophy of environmental zoning should aim at creating areas where appropriate groups of uses can coexist in a suitable setting. This attitude has been entertainingly expressed by Nan Fairbrother in the following words: "Single land uses seldom create an environment any more than separate piles of butter and sugar and flour constitute a cake; for, like a cake, an environment is a complicated whole created by skillful blending and fusing of suitable raw materials." Ideally, zoning should be considered not as a restrictive process but as a constructive one; its goal should be to integrate different types of land uses that would interplay interestingly in planned environments.

Thus, long-term planning for the countryside around large urban centers must consider not only the present condition and uses of the land, but also its potentialities. We need comprehensive maps of what has been termed "land capability classes" which emphasize, not existing uses, but rather the potentialities inherent in the soil, topography, climate, relation to water, etc. Such knowledge might suggest ways of developing food production and other desirable uses of the land in certain urban areas. This in turn would contribute a sense of spacious and orderly planning which proclaims that society can manage its environment in a way compatible with the welfare of both humankind and the earth.

The last example of priority I shall consider is one almost peculiar to the United States. I was sensitized to it the very first evening I spent in this country and I am still impressed with it after all these years—namely the urgency for a better management of waterfronts both within and outside urban areas.

As a twenty-three-year-old immigrant I arrived in New Brunswick, New Jersey, in early October 1924 and took a room in a small hotel one block removed from the Raritan River. I naturally felt somewhat forlorn on this first day in a foreign land, but being eager to investigate my new surroundings I decided to take a long walk along the river, just as I would have done in any European town. To my great surprise and disappointment, however, I discovered that it was practically impossible to reach the banks of the Raritan anywhere near my hotel, let alone walk along a river path, and I therefore had to be satisfied with watching an endless stream of motor cars crossing the bridge over the Raritan.

One of my first long trips in the United States was to Omaha, Nebraska. I had read with great delight that Omaha was located on the Missouri, a river that I was eager to know because it had nourished my imaginings about the American continent ever since the days of my youth in France. But once again I was disappointed because the riverfronts in Omaha were then, and probably still are, occupied by highways, railroad yards, and industrial buildings, which made them essentially inaccessible to pedestrians within the city limits. I left Omaha without having really experienced the Missouri.

I have lived in New York City for the largest part of my adult life—first on Riverside Drive close to the Hudson River, then for a short time not far from the Brooklyn Bridge, then in Greenwich Village close to the lower Hudson and during the past thirty years in mid-Manhattan, always within a very few blocks of the East River. Now that I have visited many of the most famous and largest cities of the world, I cannot think of any one of them that equals New York City for the diversity of its waterways and waterfronts. New York City has 578 miles of waterfront along the Atlantic Ocean, the Hudson River, East River, and the Harlem River! Herman Melville was so impressed by the contributions of the waterscapes to life in Manhattan that he referred to the subject in the first page of *Moby Dick.* He described the "crowds of water-gazers" that assembled then at the Battery; he marveled at the fact that "right and left, the streets take you waterward," where people of all social classes could vicariously enjoy the whole world, watching the ships that reached the city over rivers and oceans.

The New York City waterfronts have lost much of their romantic appeal since Melville's days and most of them are almost inaccessible to pedestrians. It takes much effort to make contact with the Atlantic Ocean from Battery Park or Staten Island; and to approach the great bridges—including the Brooklyn Bridge—that link Manhattan to the surrounding boroughs and to New Jersey. Furthermore, it has become all but impossible to escape from the fumes and noise of automobile traffic anywhere along the Hudson, the East River and the Harlem River. Much the same can be said of waterways and waterfronts in many other American cities. There is probably no other part of the world in which cities have been so generously endowed with natural waterscapes but also no country where urban waterfronts have been so grossly mismanaged and spoiled.

Excuses for the neglect of waterfronts in the United States are easy to find in history. Rivers were at first the chief and most convenient means of access to the cities of the new continent, so that their banks came to be used for commercial and industrial purposes. Furthermore, most river valleys were infested with mosquitoes that transmitted malaria, so that the tendency was to establish residential sections as far as possible from the rivers and lake shores. Until recently, indeed, there was a tradition in much of America that waterfronts were only for poor working people and for bums.

Conditions have now changed; commerce is no longer dependent on waterways; mosquitoes can be controlled and malaria has been practically eradicated. The quality of life in American cities could be greatly enhanced simply by making the bodies of water and the waterfronts centers of attraction for city dwellers and visitors. Few are the aspects of nature that offer as great a diversity of environments and as wide a range of distractions as do the rivers, lakes or oceanfronts on which most cities are located.

Even though many rivers and lakes are now polluted this is no reason for despair because, in most cases, environmental damage is reversible. The recuperative power of nature is so great that, wherever steps have been taken to *prevent further pollution* from industrial and domestic sources, the bodies of water have cleaned themselves by their own natural mechanisms as have for example, Lake Washington in Seattle, the Willamette River in Oregon and Jamaica Bay in New York City.

Pollution control is only one aspect of a complete policy of environmental enhancement. Another aspect is to take advantage of waterways and waterfronts for creating pleasant surroundings readily available to city dwellers. Such attractions will become of increasing importance as travel difficulties make it necessary for more and more people to spend leisure time close to their places of residence.

European cities have a long tradition of transforming the bodies of water—even dull and minor ones—on which they are located into environments that greatly add to the charm and diversity of urban life. Painters of the Impressionist school have left many records of festive occasions associated with small rivers and lakes. Naturally, several American cities have also succeeded in taking advantage of their waterfronts for public enjoyment. A famous example is the spectacular Pacific shore in San Francisco. Furthermore, highly desirable achievements are possible even where natural conditions are not so obviously favorable. During my association with Harvard University Medical School thirty years ago, I spent happy hours walking along the banks of that part of the Charles River that had been landscaped by Olmsted. Even a small unspectacular body of water, such as the one in San Antonio, Texas, can make an important contribution to urban life provided it is well integrated with the cityscape, unspoiled by automobile traffic, and of easy access to pedestrians. Finally, and probably most important for urban life, the waterscape must not only be attractive from the scenic point of view, but also be designed in such a way that it can serve as a stage inviting people to engage in pleasurable activities—fishing, boating, picnicking, singing, dancing, acting, and also daydreaming.

Economists will object that improving waterways and waterfronts purely for the sake of enjoyment cannot be regarded as a serious compelling urban priority—but neither was the creation of the Anthropological Museum in Mexico City, nor the archeological program in Communist China, nor the rebuilding of ancient Warsaw and of the Hamburg Opera

House at the end of the Second World War. The violent objections to the plans for Central Park at the time of its creation provide a lesson in this regard. There was no compelling reason for such an ambitious undertaking as the creation of Central Park at the time when the city still occupied only lower Manhattan. And yet, how much poorer a place New York City would be without Central Park in Manhattan and Prospect Park in Brooklyn.

Enriching urban life with the daily satisfactions that can be provided by imaginative use of waterways and waterfronts has an immense value, even though it is one that cannot be measured in dollars and cents. If economists do not know how to measure this value, and to incorporate it in their formulation of social priorities, then economics is even more dismal a science than it has been said to be.

As I write these lines, a violent controversy is raging in New York City concerning the relative merits of using public funds to create Westway, an underground highway covered by a public park along the lower Hudson, or of using these funds to improve the public transportation system throughout the city. I know that New York City subways and buses should be improved—even if this does contribute still further to the degradation of the core city by encouraging people to move out of it, but I am even more convinced that creating a park with easy access to the lower Hudson would be an immense contribution to the enjoyment of life in New York City for both inhabitants and visitors. How do you put a price on the experience of the "crowds of water-gazers" in Melville's *Moby Dick* who became "fixed in ocean reveries" as they watched the great ships sailing up and down the Hudson.

Refusing to regard as a social priority the improvement of urban waterfronts seems to me a lamentable expression of a narrow view of life which is perhaps the most disturbing threat to America's greatness today and in the future.

DAYDREAMING ABOUT THE FUTURE

I am eighty years old as I write these lines. Although I have suffered more than an average share of organic diseases, I am still vigorous enough not only to resent many aspects of modern civilization but more importantly to enjoy the world and have faith in its future.

Having lived so long in different places under various circumstances, I have become convinced that resiliency is a universal attribute of all living organisms—from natural ecosystems to individual human beings; it is also one of the most important. In living organisms, resiliency implies the ability both to recover from traumatic experiences and to create new values during the very process of recovery.

I believe also that we can improve human life and the environment, not by attempting an impossible return to the hypothetical world of Rousseau's noble savage, but by social and technological innovations that will reveal and activate potentialities of human nature and of the earth hitherto unexpressed. It is because our approaches to this problem are so clumsy that I despair but my judgment of history and of our potentialities gives me enough hope to have made me adopt the phrase "The Despairing Optimist" as title for the column I used to publish in the *American Scholar.*

I know that a very large percentage of contemporary enlightened people feel that any form of optimism is practically incompatible with the realities of our times. And I must acknowledge that pessimism overshadows optimism in my own view of a few social problems, in particular with regard to nuclear warfare and to unemployment among young people, as I mentioned in the preceding section.

In contrast, I am rather optimistic concerning the social priorities also mentioned in the preceding section as well as concerning resource and technical difficulties that industrial societies are experiencing now or will experience in the future. As discussed earlier, we are developing skills that enable us to anticipate some of the likely consequences of natural events and of the courses of action we envisage. Furthermore, scientific knowledge enables us to learn how to solve most of the practical problems of the modern world—from shortages of food or energy, to environmental degradation and probably even overpopulation.

Most of the statements I have made so far in this book refer to issues concerning which I have reached conclusions which have for me almost the strength of convictions. I shall now consider some trends which appear to be emerging in modern societies and that I am observing with even more wishful thinking than I had displayed in the earlier parts of this text.

Most important in the long run, probably, are the scientific discoveries recently made concerning the human brain and its influence on the state of the body in health and in disease. I have selected the following examples only to illustrate how wide the range of advances is in this field of research.

It has been known for a few decades that, although the right and left hemispheres of the cerebrum appear anatomically identical, they serve very different functions, as if we were endowed with two different minds. In general, the left hemisphere appears dominant in ordinary life because it controls such attributes as ability to speak, to engage in analytical thought and to develop the skills required for calculations. In contrast, the right hemisphere is involved in processes concerned with feelings, with artistic creation and appreciation, and with gestalt perceptions. It is as though, in most of us, the left hemisphere is responsible for the mundane practical jobs of life, leaving the right hemisphere free for dealing with other less critical issues. Although functional differences between the left and right brain are real in all normal persons, very recent discoveries have revealed that the situation is more complex than used to be believed. For one thing, the division of labors between left and right hemisphere is not precisely defined. Furthermore, in children who have suffered partial or even total destruction of one of the hemispheres as a result of disease, accident or surgery, the functions normally served by the lost hemisphere can be progressively taken over by the other one. There is even some evidence that a slight measure of functional recovery can take place also in adults. Some neurophysiologists have gone so far as to state that "all intellectual functions of the brain and all motor functions can be properly performed by one hemisphere alone; and that it makes no difference which hemisphere it is." Although this is certainly an overstatement, there is reason to believe that, at least early in life, both hemispheres have almost the same potentialities and that their functional differences are the result of later specializations, probably occurring around the age of two or three, the causes and advantages of which have not yet been explained.

In 1978, the Japanese neurophysiologist Tadanobu Tsunoda claimed in his book *The Japanese Brain: Brain Function and East-West Culture* that the language one learns as a child influences the way in which the brain's right and left hemispheres develop their special talents. The differences between Western and Eastern minds might thus in part be manifestations of how patterns of brain organization, influenced by external agencies, determine specialized ways of dealing with the world—from spatial and verbal activities to musical or artistic gifts.

One of the great neurological puzzles is the storage of memory in the brain. Many theories have been formulated to locate memories in specific areas and in certain chemical components of the brain, but all of these hypotheses have been shown to be inadequate. There is presently emerging, however, an entirely new theory of biological memory—more generally of neural information—which has a special appeal because it amounts to a theory of the very nature of mind. It is called the "hologramic theory." My only justification for mentioning a topic of which I have no scientific knowledge is to make the reader of this book receptive to radically new concepts of the mind that appear to me likely to become extremely important in the near future.

In optics, holograms encode messages carried by waves. According to specialists in this field "holograms of all types share in common the fact that they encode information about a property of waves known as *phase.*" Odd as it may sound, a phase has no definite size or absolute mass; in fact, it was virtually unknowable until the development of the branch of optics known as holography. To reconstruct phase, which is possible with a hologram, is to regenerate a wave's shape and thus recreate any message or image that the original wave communicated to the recording and storage medium. According to the hologramic theory, the brain stores what we call "the mind" in the form of codes of wave phase.

An entirely different line of investigation has revealed that the brain produces a multiplicity of hitherto unknown hormones, referred to as endorphins and enkephalins. The word endorphin was first introduced to denote a particular hormone produced by some part of the brain, which has effects similar to those of the drug morphine. There is evidence that the secretion of endorphin is increased under certain stressful conditions—perhaps even at the time of death—and thus decreases the perception of pain. This endorphin is only one of the many hormones that have recently been shown to affect one or the other functions of the brain, and therefore the perceptions of the mind.

It has long been known, of course, that the way we respond to environments, to human contacts, to various forms of disease or indeed to any experience is greatly affected by our state of mind. And everyone knows that almost anything that affects the body also influences the state of mind. Thus, we may be approaching the day when it is possible to understand, and perhaps control to some extent, the mechanisms through which our brain interplays with our body and conditions the responses our total organism makes to surroundings and events. In fact, the body-mind relationship is likely to become one of the most important fields of biology and medicine in the near future. The main trouble with the modern sciences of biology and medicine is that they are too one-sided. They will become truly scientific only when they have committed themselves to the doctrine that, in all aspects of life, particularly in human life, the body and the mind are intimately linked in all the manifestations of health and disease.

Schumacher's phrase, "Small is beautiful," achieved rapid and widespread popularity probably because it corresponds to some deep human longing—a quasi-universal desire in the industrialized world to escape from the huge social and technological megamachines that control ever-increasing aspects of our lives. Regions are preferred to large nation-states, moderate-size towns to anonymous urban agglomerations, individual enterprises to assembly lines, boutiques to shopping centers or large department stores. Whether decentralization and differentiation are really the ways of the future remains to be proven, but recent technological trends and the economics of modern societies seem to be in harmony with the human inclination symbolized by the phrase "small is beautiful." In private enterprises as well as in government agencies, the costs of administration and coordination escalate disproportionately with size. Many institutions have reached the point of diminishing returns with regard to size and not a few are at the point of negative returns. Fortunately, changes occurring throughout the technological and social orders point to the possibility of making certain kinds of smallness (usually qualitatively different from those advocated by Schumacher and his followers) compatible with technological and social success. Ever since the beginning of the Industrial Revolution, there has been a tendency to concentrate the manufacture of most products in larger and larger industrial units and to distribute these products over longer and longer distances. This policy, which has been economically profitable so far, is likely to be reversed as a result of changes brought about by the communication revolution and by new attitudes toward work.

The cost of physical transport will almost certainly rise faster than other costs, not only because of shortages of fossil fuels but also because there does not seem to be much chance for further technical improvement in the traditional methods of transportation. In contrast, the systems of electronic communication over the air are constantly becoming cheaper, more diversified and more practical. The differences between physical transport and electronic communication will therefore create economic incentives, on the one hand for dispersing physical manufacture and on the other hand for linking electronically the dispersed operations to central facilities responsible for design and control.

The advent of microprocessor technology has already begun to convert brute *machines* into more human *tools.* Whereas, in the early phases of the Industrial Revolution, the factory worker was the servant of the machine, he tends to become the manager or even the master of tools in modern technologies. In other words, the industrial jig is giving the factory worker a new chance to become a craftsman.

Small computers and other forms of sophisticated tools may lead to a do-it-yourself high technology carried out in small industrial units—even at home in certain cases. The sophisticated tools will of course have to be grouped at convenient locations so that they can be appropriately serviced, but people will have a much better chance than they had during the past two centuries to plan and govern their work as they do their free time.

In addition to their decentralizing effects on the technical operations of industry, the advances in communication technology and in other aspects of microelectronics will make it possible for both private and governmental administrative enterprises to become more dispersed and to evolve into federations of small semiautonomous units.

Labor policies are likely to undergo even more profound changes. The general policy at the present time is that a few professional people charge *fees* for work done whereas most other people—whether professionals or ordinary employees—are paid *wages* for time spent on the job. As problems of employment become more costly and complex, employers may tend to contract out as much work as possible to individuals or groups. In any case, the number of manufacturing and office jobs will probably decrease as happened to jobs on the farm following the mechanization of agriculture.

As more and more people become essentially self-employed they will tend to have specialist rather than organizational careers; when in need of help, they will look for support not from employers but from professional or trade associations that will replace labor unions. Contracting out manufacturing and administrative operations will become more attractive as electronic techniques make it easier to control both quantity and quality in these operations. This will be particularly true for very sophisticated technologies.

Changes had begun to take place in the attitude of the labor force and consequently in employment policies even before the microelectronic era. The best known of these changes occurred in the operations of the assembly lines of motor car plants and other highly automated industries. Efficiency experts have tried to apply engineering mentality to industrial workers, but human beings refuse to be engineered; the more automated the plant is, the greater the amount of absenteeism. For this reason much experimentation is now focused on changing the organization of factory work. An early example was the Kalmar concept introduced by the Volvo Company in the production of its automobiles. According to this concept cars are being assembled, not on the traditional assembly line, but by small groups of workers who decide themselves on their own system of cooperation and on the rhythm of their work. The most remarkable aspect of the Kalmar plan is that it started from personal initiatives among the workers and progressively evolved as they learned to work together in small groups.

The trend toward decentralization will be reinforced by subtle, but important psychological factors. In order to be successful the use of authority will have to become increasingly personal. Fewer and fewer people will be willing to *obey;* the most to expect of them will be that they *agree* to instructions. Administrators will not be effective in leading or influencing people in their organizations unless they relate to them personally. Such personal relationships put a limit on the number of people that make an effective operational unit. The proper size may range up to a few hundred people, but certainly not many thousand. Bureaucracy will undoubtedly survive but it would take a more human and less anonymous face if it could be largely contained within communities of a reasonable size.

There may be a biological basis to the optimal size of human groups. As mentioned earlier, the human bands of the Stone Age consisted at the most of a few hundred people, as did agricultural villages throughout history. When our contemporaries dream of going back to the land, they may be fantasizing about the charms and ease of country life, but they are also motivated by some deep biological longing for an appropriate size of human community, determined in the course of evolutionary development.

The diseconomics of size are particularly striking in huge urban agglomerations. Just as medieval towns usually did not exceed 50,000 in population because this was the largest number of people that could be fed by farms located within a distance reasonable for wagon travel, so the difficulties of administration and of waste disposal will probably limit the further growth of modern cities. It is likely, furthermore, that city size will also be limited by the cost of transport of food and by the uncertainties in supplies resulting from strikes and international conflicts. The food shortages experienced on the Atlantic coast in the United States after only two weeks of a partial truckers' strike in June 1979 are a warning of the dangers inherent in complete dependence on food which has to be hauled over long distances. The phenomenal increase in the price of vegetables and of fruits, as well as the decrease in their gustative quality may be among the first factors limiting the size of cities and also encouraging a renewal of food production in urban areas. I am aware of the ordeals of traditional methods of food production, and am not advocating a return to these old ways. But I also believe that modern knowledge will make it possible to develop practical methods for the production of certain kinds of food in suitable urban areas. A few of these methods are among the shows in the new Disneyland now being developed in Florida but there are others which have reached a state of practical application. For example, in a project developed by the British Glass House Crop Research Institute, the plants are grown in polyethylene film gullies with a small, one degree slope. The nutrient solution flows by gravity over the rootmass of the plants, goes to a catch tank, from which it is pumped back to the initial point and then recirculated. This method is being used in some seventy countries for the production of high cash crops such as tomatoes and cucumbers. Tomato plants twelve feet high with immense root masses are thus being produced in several places. The technique is said to be low in cost. It is relatively simple and uses very little water. It has been recently applied to the production of turf grass used for grazing, and it may soon be extended to the production of rice and wheat.

Re-creating truck farms around towns and cities, or introducing entirely new types of vegetable production would help to dispose of urban organic wastes, would avoid the high cost of transporting fresh food over long distances, and last but not least would make ripe fruits and vegetables available to the city dweller. Cottage industries using sophisticated technologies may soon permit the local production of several kinds of foodstuffs and to brand as antisocial the importation of apples, lettuce, cabbage, carrots and certain other vegetables and fruits from hundreds or thousands of miles away. The same may apply to radio and television sets, perhaps even to certain types of automobiles.

While human life is obviously influenced at each step by genetic and environmental factors, its more interesting aspects transcend simple deterministic explanations. In 1605, at the very beginning of the scientific era, Francis Bacon wrote in the *Advancement of Learning,* "The invention of the mariner's needle which giveth the direction is of no less benefit for navigation than the invention of the sails which giveth the motion." This metaphor clearly meant that progress would depend on the formulation of goals as much as on the development of techniques. Bacon certainly believed furthermore that goals are always influenced by judgments of values. His warning did not have much influence until recently because motion rather than direction has been the chief concern of those responsible for economic and technological development; but the climate of opinion is beginning to change. While bigness and speed are still the most widely accepted criteria of success, we have come to realize that etymologically the word progress means only moving forward—and for all we know perhaps on the wrong road.

The concept of progress may well be indefinable because it can refer to several different processes of change which are unrelated and some of which are incompatible. Progress can mean logical sequential advances that can be converted into computer language; it can also mean intuitive changes associated with value judgments. During much of history, the myth of eternal return has dominated human thinking about the future. Progressively, however, it has been replaced or at least supplemented by the belief that everything is continuously moving toward some omega point. While Ecclesiastes taught that there is nothing new under the sun, we are increasingly inclined to believe that our role on earth is to build the New Jerusalem. The myth of eternal return has much appeal because it provides the satisfaction of experiencing diversity while being part of eternity, but Western civilization has long been committed to a more dynamic view of life that implies continuous creation not only of new material goods but also of new knowledge and new values. The ancient Promethean myth symbolizes the belief that we are engaged in a continuous process of change to a new state that will be different from anything in the past, even though it may not be exactly what we want.

The feeling has become widespread during the past few decades, however, that the new is not necessarily preferable to the old and that our own type of civilization may not be as superior as we once believed to those we still call primitive. This is particularly apparent when we shift the emphasis from economic wealth and technologic power to the quality of life and of the environment. Increasingly also we tend to feel that, at least for human beings, some tangible link to the past is an essential ingredient of happiness. We long for an ethos that would give us, as did the myth of eternal return, both the excitement of change and the security of returning to a state and place where we feel at home.

Charles Lindbergh's life, as reported in his posthumously published *Autobiography of Values,* symbolizes how the modern world has evolved from unquestioned and uncritical fascination with sophisticated technologies to the fear that excessive dependence on them threatens fundamental human values. While on a camping trip in Kenya during his late adult life, Lindbergh became intoxicated with the sensate qualities of African life which he perceived "in the dances of the Masai, in the profligacy of the Kikuyu, in the nakedness of boys and girls. You feel these qualities in the sun on your face and the dust on your feet . . . in the yelling of the hyenas and the bark of zebras." Experiencing these sensate qualities made Lindbergh ask himself in his autobiography, "Can it be that civilization is detrimental to human progress? . . . Does civilization eventually become such an overspecialized development of the intellect, so organized and artificial, so separated from the senses that it will be incapable of continued functioning?"

Lindbergh's doubts concerning civilization and progress were the more surprising to me because I had known him in the 1930s as a colleague at the Rockefeller Institute for Medical Research where he was developing an organ perfusion pump in collaboration with Dr. Alexis Carrel. His dominant interest at that time was, along with aviation, mechanical devices to explore what he calls in his book "the mechanics of life." The *Autobiography of Values* reveals how he eventually supplemented his interest in mechanical skills with a deep concern for their social and philosophical implications. He retained an intense interest in modern science and was, for example, fascinated by space exploration, but he also became increasingly distraught at seeing technology being used for trivial and destructive ends.

Thus, Bacon at the beginning of the scientific era, and Lindbergh more than two centuries later, expressed in different words concern for one of the central problems of modern civilization. Science and technology provide us with the *means* to create almost anything we want, but the development of means without worthwhile *goals* generates at best a dreary life and may, at worst, lead to tragedy. Some of the most spectacular feats of scientific technology call to mind Captain Ahab's words in Melville's *Moby Dick:* "All my means are sane, my purpose and my goals mad." As I stated above, however, the demonic force is not scientific technology itself, but our propensity to consider means as ends—an attitude symbolized by the fact that we measure success by the gross national product rather than by the quality of life and of the environment.

While it is easy to agree that the goal of technology should be improving the quality of life and of the environment, rather than simply increasing the quantity of things produced, it is probably impossible to imagine qualitative changes that everyone would judge desirable. The very word desirable implies value judgments which are largely outside the realm of scientific inquiry because they are made on the basis of individual tastes and prejudices.

Scholars whom I greatly admire believe for example that we should orient much scientific research and development toward the colonization of space. Many American scientists and technologists have formulated the theoretical and practical merits of this enterprise and one of my French acquaintances, a humanist, sees in it an indispensable need for the continued growth of Civilization because all prior civilizations have been limited by the finiteness of our planet. I listen to them, interested but unconvinced. In my opinion we may dream of stars and of other worlds; we may even engage in flirtations with them, but we are part of the earth and can survive only by remaining bound to it—as by an umbilical cord. The people in a space laboratory, whether American or Russian, could not survive if they were really placed in the atmosphere of space! Present spaceships and their atmosphere are designed to enable a kind of life and activities very similar to those that prevail on earth. We may develop the technology for building space colonies, but these could become really habitable for human beings only if we could establish in them a physicochemical and biological environment essentially identical with that of the earth. In order to be suitable as a human habitat, a space colony would have to become a fully integrated ecosystem—including organisms ranging from photosynthetic plants that generate a breathable atmosphere, to the immense diversity of microbial species that recycle organic matter. In my opinion, perhaps based on prejudices encrusted by my years, creating in space such a complex and self-sufficient ecosystem suitable for human beings is all but impossible and therefore makes space colonization a technological goal of questionable value. This, however, does not decrease in any way the importance and interest of space sciences.

The fact that all human beings have the same fundamental needs and the same fundamental patterns of behavior seems to imply that it would be easy a priori to design on earth utopias capable of providing universal happiness, but this is unrealistic. Utopias are stillborn or soon expire because environmental conditions and human aspirations are forever changing. Furthermore, the very concept of utopia assumes that we know most of what is needed for human happiness, whereas it is certain that the word happiness means very different things for different persons, even of the same age and social group. For some it is hiking alone in the wilderness and for others being part of the crowd on New Year's Eve around 42nd Street and Times Square.

The fallacy of the utopian assumption has been recently illustrated by the failure of extremely modern architectural developments to provide satisfactory dwelling and working facilities for their occupants. A tragic example is the huge Pruitt-Igoe housing project in Saint Louis, which had been highly praised for its design and construction in the late 1950s yet had to be dynamited in the early 1970s, less than twenty years after it had received prizes for providing what were then assumed to be ideal facilities for middle-class people. The planners who had designed the project apparently had little understanding of the ways of life and tastes of the population for whom it was built, the result being general dissatisfaction of the occupants, uncontrollable vandalism and insecurity in the hallways. Planners and architects rarely live in the types of settlements they design and build for their clients, whether rich or poor.

Some of the value judgments involved in the formulation of technological goals have their basis in universal characteristics of human nature, but others are as diversified as are the cultural traditions and aspirations of the various groups of humankind.

Since all human beings acquired their genetic identity from the same distant progenitors and since it is this interplay with the various characteristics of our planet which have shaped our civilizations, we can remain human only by formulating goals which are compatible with our biological and earthly history. The atmosphere of the earth which we breathe and the cultural atmosphere in which we move must be consonant with our ancient past. Happiness and the art of living depend, to a large extent, on satisfying very ancient needs in a modern context.

Although the art of living has been expressed in many different forms, most of these have in common certain factors that probably derive from the characteristics of the kind of scenery in which humankind emerged—a land of hills and valleys, with streams and lakes, a diversity of animals and plants, alternating rainy and dry seasons associated with growing and resting periods of vegetation. For thousands of years, diversified environments of this type have inspired the themes of mythology and our images of paradise on earth. Countless stories and paintings have depicted pastures and croplands cared for by farmers, trees under which shepherds tend grazing animals, groves with bodies of water or fountains where young people engage in the games of love, and adults meditate or philosophize. We may identify Beduins and Tuaregs in our minds with vast treeless expanses of sand, but in their own thoughts they probably often long for the oasis.

The greatest difficulties in formulating technological goals come from the fact that the human invariants have been expressed in the many different forms that we call civilization. Names such as Sumer, Mesopotamia, Egypt, India, China, Greece, Islam, Europe with its many faces—from the Dark Ages and the Middle Ages to the Renaissance, the Enlightenment and the Technological Age—call to mind innumerable attempts at desirable ways of life, each implying special goals for the technology of the time.

Whatever the nature of the technological goals, all of them imply the expenditure of energy. Harnessing fire, approximately one million years ago, was humankind's first technological leap, symbolized by the Prometheus legend. Human life has been organized around fire for so long that the flame has deep and mystic undertones for all human beings. In satellite pictures of the earth, taken from its dark side at night, the most dominant feature in our times is the presence of fire. Even Africa is constantly ablaze with pinpoints of light from brush fires, from cities or from oil refineries.

Until the Industrial Revolution, all technologies were powered exclusively by muscle, wood, wind or waterfalls, which derived their energy from the sun. Even coal, petroleum and natural gas to which industrial societies have now become addicted have their origin indirectly in the sun since they derive from the photosynthetic activities of plants during geological periods.

Thus, all the great civilizations of the past were powered by the sun, directly or indirectly. The Greek philosophers, Leonardo da Vinci, Michelangelo, Shakespeare, Newton, Goethe lived before the fossil fuel era. Baudelaire, Picasso and Einstein would almost certainly have been just as creative if coal and petroleum had not been used during their times. In fact, all aspects of the arts and sciences, all manifestations of love, joy and enthusiasm have been expressed for thousands of years by people who had no access to fossil fuels.

The province of France called La Beauce has a highly productive agricultural soil and is crowned by one of the world's most famous monuments—the cathedral of Chartres. This land has been made fertile by the labor of peasants who have cultivated it for several millennia, without fossil fuels. Similar statements could be made about all the humanized landscapes of the world. New England and the Pennsylvania Dutch country are examples on the American continent. Thus, human interventions into nature have often resulted in the creation, not only of new environments, but also of artistic and spiritual values as in the Chartres cathedral. The timbre of bells, the daring of the architecture, the mystical quality of the light that filters through the stained glass windows, are products of an extremely complex thirteenth-century technology. In many different ways this technology was dependent on the use of solar energy which thus helped human beings to experience the cosmic order through their sense organs, and transport the human spirit beyond metal, stone and glass. Long before the fossil fuel age, many other technologies have similarly contributed to our understanding and enjoyment of the rest of creation.

We have been children of the sun since the beginning of time. As the Italian proverb goes, "We cannot all live on the piazza, but everyone may enjoy the sun." If there has been acceleration in the growth of knowledge and in the production of material goods during recent centuries, it is because we have used reserves of solar energy accumulated in the form of trees, coal, petroleum, natural gas. All forms of knowledge and of art are hymns to the sun and even at night, when the moon shines in the sky, the poetical quality of moonlight is itself a reflection of the sun.

The mining of the solar energy stored in the form of fossil fuels will eventually come to an end. Our awareness of this fact accounts in large part for our worries about the future, but we should derive comfort from the fact that many civilizations have reached immense heights and have lasted for centuries and even millennia, long before fossil or nuclear fuels were available—an era that will probably be remembered in the future as only a minor episode in the human adventure. Now that our knowledge is much greater, we can go much further toward creating new forms of civilization by using the endlessly renewable forms of energy derived from the sun.

At the present time solar energy is used chiefly in the form of wood. In 1870 wood accounted for some seventy percent of the total energy used in the United States; this percentage reached its lowest level, approximately two percent, around 1975 but it has begun to increase and is now of the order of three to four percent, which, surprising as it may sound, is more than the total amount of energy produced at present by nuclear reactors. It has been suggested that the use of wood and other forms of the biomass will continue to increase until the end of this century and may then account for some eight percent of the total energy requirements of the United States; but this will almost certainly be the upper possible limit.

Solar energy can be utilized of course by many techniques other than those involving the use of the biomass but solar technologies applicable on a large scale are yet to be developed in a practical form and probably will not meet the needs of industrial societies until sometime late in the next century. Fossil fuels and nuclear reactors will therefore remain the main sources of energy for several decades. In view of the well-known dangers associated with these sources, there are legitimate reasons for pessimism about the long-range future of technological civilization. In my opinion, however, the prospects are hopeful if we look at solar energy from a more futuristic point of view.

Even taking into consideration its use by all forms of vegetation, only a minute percentage of the solar radiation intercepted by the earth is utilized for human purposes and other forms of life. I do not have the kind of knowledge that is needed to convert this unused solar energy into a usable form, but I have enough faith in the human intellect to believe that appropriate sciences and technologies can be developed within the next hundred years to make us completely independent of fossil fuels.

Paradoxically, I see our greatest dangers for centuries to come not in shortages of energy but in excessive use of it, either for our own individual lives or for the manipulation of the environment. Judging from the behavior of our species since the Industrial Revolution (which should more properly be called the Fossil Fuel Revolution), it is to be feared that we shall inhibit some aspects of human development and spoil the quality of the earth by yielding to the physical and intellectual laziness into which we shall be tempted once energy is so abundant that we can use it thoughtlessly.

Whatever the origin and form of energy, excessive use of it is always dangerous if we are not sufficiently concerned with its consequences, now and for the distant future. In as far as they are influenced by human activities, the quality of our lives and the welfare of the earth will be determined, not by technical difficulties or potentialities, but by judgments of value.

The ethical difficulties involved in the applications of scientific technology were poignantly expressed in a letter that Charles Lindbergh wrote in 1970 to Congressman Emilio Q. Daddario, who had invited him to become part of a panel dealing with the organization of the Office of Technology Assessment. Lindbergh refused to join the panel but stated in his letter that he was deeply interested in the congressman's new approach to scientific research and development through consideration of its effect on the future welfare of humankind.

"The human intellect is becoming aware of the vulnerabilities that accompany its power . . . and that to avoid self-destruction it must exercise control over its accumulated knowledge. . . . We can lay down certain principles, one of them being that man must place more value on the *human lifestream* than on himself as an individual . . . that his salvation and immortality lie in it [the lifestream] rather than in himself. Possibly this will involve an intellectual religion rooting into the intuitive religions of the past [italics mine]."

The question raised by Lindbergh in his letter to Congressman Daddario brings up one of the most difficult dilemmas of humankind, perhaps the ultimate dilemma, namely the relative importance of the individual person and of the community in the management of human affairs.

There have been many societies, especially in the past, in which the community took precedence over the individual person. In a detailed account of his life around 1880, for example, a Navajo called "Left Handed" recounted how he and his people had little latitude for personal choice. Indeed, to take up any life except that of their fathers seemed not to have occurred to the Navajos of his time. But everywhere they went, they moved among kinsmen and the kinsmen of kinsmen; they lived in a world of people bound to one another by ties of blood, clanship and *obligation.*

A similar communal attitude is still vigorously alive today among the Hutterites, a Christian population that originated in Europe and has now established 230 small colonies in the United States and Canada. The goal of child rearing in these colonies is voluntary submission of the will of the adult person to the good and the will of the community. As a result of rigorous training and schooling few youngsters care to leave the community even though life in it demands much harder work and more strict religious discipline than what is expected in the normal society with which the Hutterites are in direct contact.

Almost everywhere else, however, social evolution has given an ever-increasing importance to individuality, making the individual person the ultimate unit of value. "The man who is aware of himself is hence forward independent. . . . He alone lives. . . . Once conform, once do what other people do because they do it, and a lethargy steals over all the finer nerves and faculties of the soul." These words of Virginia Woolf beautifully express the rich value of individuality, but they fail to warn the reader that extreme emphasis on the rights of the individual person may create great dangers for our survival as a species. In the words of the psychologist Robert Coles, "The self is our guide, our standard—those psychological 'needs' we experience, those psychological 'passages' through which we journey, those 'emotions' we boastfully proclaim to each other." As members of the "ME" generation, many contemporary persons of all ages, but perhaps especially among the young, try to find a substitute for boredom in their own individual problems. The line attributed to the French king Louis XIV, "L'État, c'est moi" (I am the state), may have had some political merit at one time, but its equivalent in our private lives certainly leads to disasters especially when it is adopted by millions of persons within a given society.

In the final analysis, the welfare of Humankind may well depend upon our ability to create the equivalent of the tribal unity that existed at the beginning of the human adventure, while continuing to nurture the individual diversity which is essential for the further development of civilization. We should aim at some form of political unification of humankind, but global unity will be viable only if it is compatible with the cultivation of diversity and of pluralism in our habits, tastes and aspiration.

This is not the best of times, but it is nevertheless a time for celebration because, even though we realize our insignificance as parts of the cosmos and as individual members of the human family, we know that each one of us can develop a persona which is unique, yet remains part of the cosmic and human order of things. Human beings have been and remain uniquely creative because they are able to integrate the pessimism of intelligence with the optimism of *Will.*

ENVOI

The immense herds of animals that roam over the parts of East Africa where humankind is assumed to have emerged provide some of the most exciting spectacles of nature. They symbolize the power of life to create a seemingly endless diversity of species, each exquisitely adapted to a particular habitat and a particular pattern of behavior. Beautiful as these animals appear to us, however, and admirably adapted as they are to the African savanna provided they can elude their human predators, they all have the limitation of being prisoners of Darwinian evolution which is irreversible, and which determines where and how they must live.

Human beings, in contrast, are blessed with the freedom and flexibility of social evolution which is almost completely reversible. While remaining members of the biological species *Homo sapiens* we have experienced many different ways of life and some of us continue today as hunter-gatherers, as pastoralists, as farmers, as sailors, as artists, as factory workers or as reclusive scholars. We have been members of roving bands, of sedentary villages, of towns or cities, or of cloistered religious or scholarly communities.

Throughout history and also prehistory, we have had the freedom to choose our course, to change direction and even to retrace our steps in order to reach the goals we have selected. The deterministic future operates in human life as it does in other forms of life, but it has continuously and increasingly been supplemented by the willed future based on human values and aspirations.

Like other human beings at all stages of prehistory and history, we are still on the way. We constantly renew ourselves by moving on, to new places and new experiences. Wherever human beings are involved, trend is never destiny because life starts anew, for them, with each sunrise. *Demain, tout recommence.*

We may differ in our tastes and goals; we may even despise much of what we see around us, but most of us would join in Thoreau's clarion call in *Walden:* "I do not propose to write an Ode to Dejection, but to brag as lustily as Chanticleer in the morning, standing on his roof, if only to wake my neighbors up"—for a Celebration of Life.

INDEX

Adams, Henry, 197
Adams, James Truslow, 220
Adaptation, 73, 86, 179–180
  creative, 150, 182–192
  future, 145–151, 182–192
  to population and space problems, 95–119
Adaptive interventions, solutions vs., 230
Advancement of Learning (Bacon), 243
Adventure, craving for, 51, 60, 74, 80, 206–207
Age of Uncertainty, The (Galbraith), 201, 202
Aggression, 5, 13, 15, 16
Agriculture and food, 15–16, 22, 39, 41, 85, 94–95, 191–192, 198
  costs of, 95, 241–242
  energy use in, 158, 172, 175, 178–179
  in growth of cities, 205
  in modern village planning, 213, 214–215
  natural ecosystems altered by, 158–160
  in the Netherlands, 96, 97, 99, 102, 104–105, 106
  nutritional habits and, 29–30, 56–57, 142
  in urban areas, 95, 230, 242
Altruism, 5, 13
Aluminum:
  production and use of, 165, 166, 167
  as waste product, 169, 170
Animals:
  in fables, symbolic significance of, 11, 66–67
  human beings as, 9–14, 17–19, 74
  natural habitats of, 18–19, 74
  social diversity of, 14
Animism, persistence of, 45
Architecture, 16, 57–61, 95, 114, 246
  behavior and development affected by, 52–56, 58–61, 70
  form vs. function in, 58–61
  free-standing houses in, 177–178
  Gothic, 59, 131–133
  human need for mysterious in, 60–61
  Romanesque, 131, 132
  skyscraper, 57–58, 112–113, 115, 117, 131
Arts, 20, 21, 30, 31, 196, 247
  biological utility vs., 74–75
Auden, W. H., 201
August 1914 (Solzhenitsyn), 123–124
Augustus, 133–134, 135
Aurelius, Marcus, 134
Autobiography of Values (Lindbergh), 244
Aveyron boy (wild child), 25, 27

Bacon, Francis, 217, 243, 244
Bariloche Foundation World Model, The, 144
Beauvais syndrome, 131–133
Ben-Dasan, Isaiah, 138
Bettelle Memorial Institute, 164
Big Bang theory, 45–46
Biological clocks, 39–43
Biological determinism, 202
  environmental, 5–6, 12, 13, 15, 48–50, 62, 71, 72, 73
  genetic, 5–6, 12–13, 15, 71, 73, 204
  social diversity as argument against, 14–17, 29–33, 62
Biological development, changes in rate of, 56–57
Biological evolution, 6, 18, 19, 20–22, 50, 68, 182, 188
Biomass, as energy source, 85, 86, 94, 149, 181, 248–249
Birth of Tragedy, The (Nietzsche), 44
Black, Max, 140
Blake, William, 113, 160, 161
Blok, Aleksandr, 124
Body-mind relationship, 236–239
  cyborgs and, 69
Bos woods, management of, 108–109
Bougainville, Louis Antoine de, 24
Brain, 30, 42, 208, 236–238
  evolution of, 19, 20, 21
  left vs. right hemisphere of, 222, 237
Breweries, wastes from, 171
British Glass House Crop Research Institute, 242
Bronx River, renovation of, 119
Brooklyn Bridge, ambivalent response evoked by, 113–114
Brown, Harrison, 166
Buffon, Georges Louis, 12

Caesar, Julius, 96, 98
Camus, Albert, 103

Index

Cannon, Walter B., 190
Carrel, Alexis, 244
Carson, Rachel, 155–156
Carter, Jimmy, 146
Chagnon, Napoleon, 204–205
Chambless, Edgar, 89
Chicago World's Fair, of 1893 vs. 1933, 197
China, People's Republic of, 87
  archeological projects in, 226, 234
  reforestation program in, 143
Churchill, Winston, 70–71
Chute, La (Camus), 103
Cities, 205–211
  advantages of life in, 206–210, 216
  artificial urban centers vs., 88–89
  country life compared to, 51, 77, 88, 205, 206–207
  diseconomics of size of, 209, 241–242
  increase in size of, 205–206, 209
Civilization, 125–126, 205
  changes in meaning of, 195–196
  civility vs., 195, 196
  climatic factors and, 50, 125
  culture vs., 124
  progress vs., 244
  uniqueness of, 95
Civilization and Climate (Huntington), 49–50
Climatic factors, human attributes shaped by, 48–50
Clinton, De Witt, 112
Club of Rome, 141, 144
Coles, Robert, 90–91, 251
"Collapse of Humanism, The" (Blok), 124
Colossus of Maroussi, The (Miller), 87–88
Coming of Age in Samoa (Mead), 5, 12
Communal experiments, 211–212, 214, 216
Communitas (Goodman), 117–118
Communities, 204–216
  behavior and development influenced by, 53–56
  enduring value of, 87–88
  global problems solved in, 83–87
  individuals vs., 53, 250–251
  optimal size of, 87, 205, 208, 215, 241
Computers, social effects of, 221–224, 239–240
Cook, James, 24
Copper, production and use of, 165–166, 167
Cortez, Hernando, 23
Cosmic rhythms, physiological processes linked to, 39–47
Cossin, Jehan, 110–111
Council on Environmental Quality, 142–144
Creativity, 11, 44, 86, 179, 230, 247
  in adaptation to the future, 150, 182–192
  choices and, 70–75
Cro-Magnon people, 11, 56, 75
  Neanderthal people replaced by, 20–21

Culture:
  civilization vs., 124
  images of humankind provided by, 61–70
Curtiss, Susan, 26–27
Cyborgs (cybernetic organisms), 69
Cyprian, Saint, 200

Daddario, Emilio Q., 250
Dance of the Tiger (Kurten), 21
Dante Alighieri, 11
Darwin, Charles, 24
Decentralization, recent trend toward, 239–241
Deforestation, 143, 148, 192
Delaruelle, Monsieur, 76–77, 79
Delta project, 106, 109
Descartes, René, 11, 102, 222
Desertification, 85
Dio Cassius, 134–135
Distant Mirror, A (Tuchman), 200
DNA (deoxyribonucleic acid), 28–29, 46, 47, 182, 183, 187, 188
Dodds, E. R., 44
Dominance, as biological trait, 42–43
Dore, R. P., 138–139
Doxiadis, C. A., 88
DuBos, Jean Baptiste, 49
Dubos, René:
  education of, 64–66, 76–77
  genetic and environmental endowment of, 6–9, 63
  in Hong Kong Territory, 3–5
  past as alive in, 46, 47
  social conditioning of, 62–65, 68, 76–79

Ecclesiastes, 243
Education, 224
  socialization and, 64–67, 68, 76–77, 228–229
Education of Henry Adams, The (Adams), 197, 220
Ehrlich, Paul, 147
Ellul, Jacques, 196, 197
Emerson, R. W., 57
Enclosure Acts, 192
Endorphins, perceptions affected by, 238
Energy, 133, 141, 142, 158, 172–182, 191, 198, 227, 236, 247–250
  conservation of, 52, 59–60, 94, 95, 173–174, 178, 214–215
  costs of, 85, 94, 95, 149, 178, 179
  local solutions to problems of, 85–86, 94
  shortages of, 140, 172–182, 230, 249
"English people, The" (Orwell), 122, 123
Environment, 37–80
  conceptual, 31–33

Environment *(cont.)*
human role in modification of, 14, 19, 22, 30, 39–41, 50–56, 70–75, 96–119, 124–126, 147–150, 191–192, 203–204, 248
importance of diversity in, 51, 72–73, 80, 108, 116, 125–126, 246–247
pessimism in contemporary views on, 140–145
(*see also* architecture; natural resources; nature)
Erdmann, Benno, 14
Evolution (*see* Biological evolution; Social evolution)

Fables in social conditioning, 11, 62, 65–68, 78
Fairbrother, Nan, 161, 231
Family:
hearth as center of, 53–54
tranquility vs. togetherness of, 52–53
weakening of, 202, 210, 228
Feedback systems, in Gaia hypothesis, 189, 190–192
"Fight and flight" response, 42
"Fitness of the Environment, The" (Henderson), 189
Five Stages of Greek Religion (Murray), 140
Fossil fuels, 142, 149, 179, 180–181, 247, 248
costs of, 85, 94, 95, 172
pollution from, 175–176
social effects in use of, 86, 172–173, 174–175
France, 76–79, 80, 126–127
design of villages in, 53–56, 89, 213–214
educational system in, 64–67, 76–77
urbanization in, 160
France, Anatole, 20
Freedom/free will, 13, 38–39, 43, 53, 70–75, 80, 208, 217
Freud, Sigmund, 12
Functionalism, 58–61, 80
Future, 70, 140–151
certainty vs. uncertainty in views on, 199–204
computer models of, 141–145
material vs. human aspects of, 198–199
One World view of, 90–91
social adaptations to, 145–151, 182–192
Future shock, 145–146, 201

Gaia concept, 188–192
Galbraith, John Kenneth, 196, 197, 201, 202
Gardens:
French vs. American, 54
as unnatural environment, 158
Gauguin, Paul, 46
Genetic engineering, 150–151, 188
Genie (abused child), 26–27
Geological Survey, U.S., 164
Gibbon, Edward, 134
Globalization:
dangers of, 86, 120
nationalism vs., 119–127
recent trends toward, 90–91, 94–95, 119, 120–121
regionalism as counter trend to, 91–95
Global problems:
as different from past problems, 140–141
need for local solutions to, 83–87
Global 2000 Report, 142–144, 163–164
GNP (Gross National Product), calculation of, 228
Gomez, Estevan, 110
Goodman, Paul and Percival, 117–118
Great Britain, 124, 157, 242
diets in, 142
New Towns in, 89, 212–213
Greeks and the Irrational, The (Dodds), 44

Hals, Franz, 102
Hamburg Opera House, rebuilding of, 226, 234–235
Happiness, 87–88
link to past as element of, 243–244
social priorities and, 226–227
utopian assumption and, 245–246
wealth and materialism vs., 195–196, 199–200, 202, 211, 216–224
Health, health care, 56–57, 108, 143, 176–177, 196, 199, 218, 225, 226
influence of brain on, 236, 238–239
Henderson, L. J., 189–190
Hippocrates, 48–49
Hologramic theory, 238
Homeostasis, in Gaia hypothesis, 190–191, 192
Homer, 31
Homo abilis, 19
Homo erectus, 19–20
Howard, Ebenezer, 231
Hudson, Henry, 111
Human beings:
Homo sapiens distinguished from, 17–19, 22, 24–28, 61, 74, 203
origin of, 19–22
as subjects vs. objects, 13
Human development, processes in, 71–74
Human nature:
environmental influences on, 5–9, 12, 13,

Human nature: *(cont.)*
15, 23–33, 39–47, 71, 203
genetic influences on, 5–6, 8–14, 15, 28–31, 39–47, 71, 203
invariants of, 28–33, 56, 87, 246, 247
Huntington, Ellsworth, 49–50
Hutterites, communal attitude of, 250–251
Huxley, Sir Julian, 12
Huxley, Thomas, 12

India, community structure in, 88
Individuality, communal attitudes vs., 53, 250–251
Industrial Revolution, 195–196, 198–199, 205, 220, 229, 239, 240
energy use in, 172, 173, 247, 249
environmental degradation due to, 147, 160
first vs. second, 222–224
Intentionality, as human trait, 6, 13–17, 37–39, 70–75, 125–126
Ireland, Republic of, national feeling in, 124
Iron:
production and use of, 165
recycling of, 170
Isolationism, local problem-solving vs., 86–87
Itard, Jean Marc-Gaspard, 25–26

Jamaica Bay, pollution of, 118–119, 148
Japan, 47–48, 218, 219–220, 228
biological development in, 57
modernization of, 137, 138–140
traditional houses in, 52
Japanese Brain, The (Tsunoda), 237
Jastrow, Robert, 46
Jesus, 135
Johnson, Herbert, 118–119
Johnson, Samuel, 38, 122, 195, 196
Johnson, Warren, 230
Joie de Vivre, La (Zola), 14
Journal of Experimental Medicine, 28–29
Jung, Carl, 12–13

Kahn, Louis, 209–210
Kalmar concept, in auto production, 241
Kansas, future of farming in, 159
Kibbutzim, 48
as communal system, 211–212, 216
Knossus, happy attitude maintained in, 87–88
Krakatoa, eruption of, 162, 163
Kropotkin, Prince Peter, 123
Krutch, Joseph Wood, 22
Kurten, Bjorn, 21

Labor policies, 205, 211, 220, 225, 227, 236
decentralization and, 240–241
La Fontaine, Jean de, 11, 65–67
La Fontaine et ses Fables (Taine), 11, 67
Land management:
in the Netherlands, 89, 96–110, 117, 131
in New York City, 96, 110–119, 131
as social priority, 230–232
as unnatural phenomena, 155–163
Language, 9, 15, 24, 27, 30, 54
in animals vs. humans, 10–11, 23
Lao Tzu, 217
Latin American World Model, The, 144
Lawns, as unnatural environment, 157, 158
Le Corbusier, 58
Leeuwenhoek, Anton van, 102
"Left Handed" (Navajo), 250
Lely, Cornelis, 104
Lenin, V. I., 123
Lerner, Max, 122–123
LeRoy, Louis, 200–201
Lichens, symbiosis in, 186
Limits to Growth, 141, 144, 163–164
Lindbergh, Charles, 244, 250
"Living substance," use of phrase, 37–38
Locke, John, 12
London, 206, 213
as assembly of villages, 89–90
pollution in, 147–148
Louis XIV, king of France, 251
Lovelock, J. E., 188–192
Lumber industry, environmental problems due to, 171

Machiavelli, Níccolò, 11
Malinche, 23
Marshall, William, 157
Mason, Otis T., 188–189
Materialism:
architecture influenced by, 57–58, 61, 112–113
limitations of, 196, 199–200, 202, 211, 217, 218, 220
spirituality vs., 133–140
Mead, Margaret, 5, 12
Melville, Herman, 233, 235, 244
Memory, biological, 39–47, 237–238
Mercury, substitutes for, 166–167
Mexico City, 206
Museum of Anthropology in, 225–226, 234
Mies van der Rohe, Ludwig, 58
Miller, Henry, 87–88
Minuit, Peter, 111
Mirabeau, Honoré, 195, 196, 197
Mobility, social effects of, 90, 92–93, 94
Moby Dick (Melville), 233, 235, 244

Mohammed, 135–136
Montesquieu, 49
Montreuil, Eudes de, 132
Morand, Paul, 90
Muddling Toward Frugality (Johnson), 230
Murray, Gilbert, 140
Mutsuhito (Meiji), 138
Mythology, 37, 41, 45, 93, 123, 243, 244, 247

National Academy of Sciences, in investigation of Harlem, 80
Nationalism, 119–127
Native Americans, 40, 62, 90–91, 93, 124
  social diversity of, 15–16
  socialization of, 23–24
Natural resources, 84, 85–86, 95, 133, 140, 142–143, 163–182
  fables and, 65–66
  optimistic view of, 164–165
  "reserves" of, 164
  social adaptation to the future and, 147–149
  social priorities and, 230–235
  substitutes for, 166–167
Nature:
  culture vs., 61, 62, 67, 68
  humanized, 19, 22, 50–56, 96–119, 155–163, 203–204, 248
  humans in symbiotic relationship with, 125
  humans set apart from, 19, 20, 22, 39, 61, 75
  resiliency of, 150, 191–192, 198, 236
Neanderthal people, replacement of, 20–21
Neighborhoods, 89–90, 115, 205, 210–211, 213
Netherlands, solution to population and space problems in, 89, 96–110, 117, 131
Net National Welfare Index, 228
New Alchemy Institute, 215, 216
New towns movement, 89, 212–213
New York City, 206, 210, 227, 231
  architecture in, 57–58, 112–113, 131
  environments improved in, 118–119, 148
  neighborhoods in, 90, 115, 208
  parks in, 158, 226, 235
  self-discovery in, 80
  solution to population and space problems in, 96, 110–119, 131
  walking in, 51, 77, 118
  waterfronts of, 232–235
Nietzsche, Friedrich, 44
Nixon, Richard, 218
Noise pollution, 227
Nuclear power, 85, 86, 141, 149, 180–181, 198, 227, 236, 248, 249

Ogallala aquifer, in irrigated agriculture, 159
On Human Nature (Wilson), 5, 13
Orwell, George, 11, 122, 123, 124, 146
Outer space, exploration of, 43, 75, 244, 245

Paley Commission on Materials Policy, 218
Paris:
  environmental conditioning in, 77–79, 80
  neighborhoods in, 90
Parr, A. L., 52–53
Pascal, Blaise, 61, 222
Passivity, 71, 73, 125–126, 144–145
Past:
  rejection of, 218
  romanticizing of, 219
  survival of, 39–47
Pasteur, Louis, 47
Paul the Apostle, 22
Peter the Great, 102
Petit Prince, Le (Saint-Exupéry), 126–127
Phaedrus (Plato), 44
Photosynthesis, 175, 187
Physiological needs, architecture and, 58–59, 60–61
Plastics, as substitutes for metals, 167
Plato, 44
Pneumonia, research on, 29, 184–185
Pocahontas, 23–24
Pollution, 84, 109–110, 171, 175–176, 226, 227
  control of, 87, 94, 118–119, 147–148, 198, 220, 234
Population, 74–75, 89–90, 147, 198, 205–206, 209
  large, accommodating of, 95–119, 131
Population Bomb, The (Ehrlich), 147
Progress, 218, 219, 221
  meanings of, 243, 244
Property, control of, as biological trait, 42–43
Pruitt-Igoe housing project, 246
Psychological factors:
  in architecture, 58–59, 60–61
  in behavior, 12–13, 31–32, 40
  decentralization reinforced by, 241
  in homogenization of human life, 90–91

Racial differences, genetic vs. sociocultural causes of, 47–48
Racism, 42, 49–50
Randstad (Ring City), management of, 107–109
Reason, occult forces vs., 39, 44–45, 60–61
Regionalism, worldwide trend to, 91–95, 122
Religious feelings, 5, 13, 20, 21, 22, 75, 205
  architecture and, 57–58, 59, 61

Religious feelings *(cont.)*
cosmic rhythms and, 44, 45–46
fringe groups and, 203, 211
material power of, 133–140
Rembrandt, 102
Rescher, Nicholas, 219
Reston, James, 201
Rivière, Mercier de la, 217–218
Roadtown (Chambless), 89
Roebling, J. A. and Washington, 114
Rolfe, John, 23–24
Roman Empire, materialism vs. spirituality in, 133–135
Roosevelt, Theodore, 146
Rothschild, Edmond de, 137
Rotterdam, development of waterfront in, 106–107, 109
Rousseau, Jean Jacques, 12, 236
Ruskin, John, 198–199
Russianness, as cultural identity, 123–124

Saint-Exupéry, Antoine de, 126–127
Santorini, eruption of, 162–163
Scientific revolutions, exaggerated effects of, 68
Seasonal patterns of behavior, 39–41, 43
Self-discovery, 71–73, 76–80
Self-realization, 71, 73
Self-sufficiency, development of, 2
Sensation, impoverishment of, 221–222, 223–224, 230
Seven Lamps of Architecture (Ruskin), 199
Sexuality, 5, 14, 30, 56–57
Shaw, George Bernard, 200
Shelters, human need for, 22, 30, 56, 85
Shinohata (Dore), 138–139
"Significance of sections on American History, The" (Turner), 94
"Significance of the frontier in American History, The" (Turner), 93
Silent Spring (Carson), 155–156
Skinner, B. F., 12
"Small is beautiful," 239
Smells, emotional response to, 46–47
Smith, John, 23
Social class, freedom in relation to, 72, 80
Social diversity, 14–17, 29–33, 62, 67–68
globalization vs., 86
in New York City, 90, 115–117, 210
Social evolution, 6, 19–22, 61, 77, 116, 251
Social institutions, 175
as employers, 211
limitations of, 202
Socialization, 17, 23–28, 62–68, 202
as acquisition of collective symbols, 32–33
education and, 64–67, 68, 76–77, 228–229
Social priorities, 224–235, 236
differences of opinion on, 225–227
Socrates, 44
Soil, microbial life in, 183, 184
Solar energy, 85–86, 94, 149, 176, 180, 191, 247–249
Solzhenitsyn, Aleksandr, 123–124
Sounds, ritual, 45, 46
"Spaceship Earth," 188
Specialization, social effects of, 95, 210, 229–230, 240
State Department, U.S., 142–144
Strangers, suspicion of, 42
Stuyvesant, Peter, 111
Suger, Abbé, 131–132
Sullivan, Louis Henry, 58
Symbiosis, 97, 125, 185–188
Synfuels, as energy source, 149

Taine, Hippolyte, 11, 67
Talleyrand, Charles-Maurice de, 93
Technology, 84, 118, 195–199, 216–224
architecture influenced by, 57, 59, 60, 112–113
dangers of, 197–199, 221–224
in development of Netherlands, 96, 98–100, 103–107, 109–110
homogenization due to, 90–91, 119–121
humanized, 229, 230
images of humankind affected by, 62, 68–70
means vs. goals and, 244–247
microelectronic, 214, 221–224, 239–240
in modernization of Japan, 138–139, 140
risks and, 229–230
as solution to resource problems, 167
Thoreau, Henry David, 217, 253
Tillich, Paul, 74
Todd, John and Nancy, 215
Tools:
humans' use of, 15, 19, 20, 21, 221
machines as, 240
Toynbee, Arnold, 50
Transportation systems, 206, 239
social effects of, 205, 207, 211, 225, 241–242
Tsunoda, Tadanobu, 237
Tuchman, Barbara, 200
Tulp, Nicholas, 102
Turgenev, Ivan, 123
Turner, Frederick Jackson, 93–94

Uncertainty, Age of, 201–203
Unemployment, 205, 211, 220, 225, 227, 236

UNEP (United Nations Environment Programme), 85, 86
United Nations, international conferences organized by (1970s), 83–85
Utopias, 245–246

Valéry, Paul, 14
Value systems, 220–221, 224, 250
  classical vs. scientific basis for, 203–204
  effects of technological success on, 220
Vancouver Habitat Conference (1976), 84, 85
Van Deen, J., 100–101
Verrazano, Giovanni da, 110
Victor (wild child), 25–26, 27
"Village of the future," 214–216, 231
Villages, 53–56, 89–90, 213–214
  cities compared to, 51, 77, 88, 205, 206–207
  global, 87–96
Visual needs, 30, 56, 117–118, 131
Voltaire, 126
Voluntary simplicity, 218
Voyageurs (coureurs des bois), 24

Walden (Thoreau), 253
Walking, rates of, 51, 77–78
Warsaw, rebuilding of, 226, 234
Washington, George, 111
Wastes, disposal of, 168–171
Water, as energy source, 85, 94, 149, 176, 181
Waterfronts, management of, 96–110, 117–119, 232–235
Wharton, Edith, 160
Whyte, William H., 72
Wild children, 24–26, 27
Wilderness, humanized landscapes vs., 155, 156
Wilson, Edward O., 5, 13
Wind, as energy source, 85, 86, 94, 100, 103, 149, 176, 181
"Wolf children," 24, 26, 27
Wood:
  as biodegradable, 169, 170
  as energy source, 172, 247, 248–249
Wooing of Earth (Dubos), 147, 162, 192
Woolf, Virginia, 251
Woolworth Building, changes of taste and, 57–58, 112
Wright, Frank Lloyd, 231

Zionism, 137–138
Zola, Émile, 14
Zoning policies, 231
Zuider Zee, draining of, 103–105, 106